Telecom
Management
for Call Centers

Telecom Management for Call Centers

A Practical Guide

Luiz Augusto de Carvalho
and
Olavo Alves Jr.

June 2011—Version 2.0

iUniverse, Inc.
Bloomington

Telecom Management for Call Centers
A Practical Guide

iUniverse books may be ordered through booksellers or by contacting:

iUniverse
1663 Liberty Drive
Bloomington, IN 47403
www.iuniverse.com
1-800-Authors (1-800-288-4677)

Because of the dynamic nature of the Internet, any web addresses or links contained in this book may have changed since publication and may no longer be valid. The views expressed in this work are solely those of the author and do not necessarily reflect the views of the publisher, and the publisher hereby disclaims any responsibility for them.

Any people depicted in stock imagery provided by Thinkstock are models, and such images are being used for illustrative purposes only.

Certain stock imagery © Thinkstock.

ISBN: 978-1-4620-5682-8 (sc)
ISBN: 978-1-4620-5683-5 (hc)
ISBN: 978-1-4620-5684-2 (ebk)

Library of Congress Control Number: 2011917340

Printed in the United States of America

iUniverse rev. date: 10/26/2011

Contents

Introduction

Our objective in writing this book was to provide the reader with a practical guide that addresses the most common issues associated with telecom management in large call centers. This book is aimed at the telecom manager, and the techniques described here are practical and easily applicable. Our focus is on the issues the telecom manager faces in his or her day-to-day work.

There is a lack of literature in this area, and the lessons learned by the professionals in the field are rarely documented and shared. This book is an effort to document this practical knowledge. Additionally, there is a tendency in the scant available literature to concentrate on technical aspects and to neglect planning and managing.

Our intention is to provide the reader with a general view about how telecom infrastructures in large call centers should be planned, priced, negotiated, and managed. In this book, we are going to discuss the following aspects of call-center operations:

- Guidelines for cost management
- Guidelines for traffic management
- Guidelines for planning a call-center infrastructure
- Methodology to map and analyze the traffic in a call center
- Guidelines for planning a transport network
- Study of cases
- Guidelines for deploying GSM (Global System for Mobile communicantions) gateways
- Guidelines for billing systems
- Billing auditing

- Guidelines for deploying dialers
- Calculations
- Traffic matrixes

We divided this book into twelve chapters, which can be read separately. Although this book's focus is on planning and managing telecom in large call centers, many concepts described are applicable not only to call centers but also to telecom structures in general.

Chapter 1: Cost Management

The management of telecom in a call center has as its main goal to guarantee connectivity between the users and the attendance sites with the desired quality of service for the least cost possible. Therefore, cost management is a fundamental part of managing telecom in call centers. Traditionally, there are four main general strategies to keep the telecom cost down:

- Pressure the service providers and hardware vendors, trying to guarantee low prices—for example, bargaining and negotiating hard.
- Enhance the internal control over usage of the services available—for example, by using billing systems.
- Increase the control over the service providers to make sure that the organization is paying only for what was really used at the agreed price for each service and is not being overcharged—for example, by using telephone bills auditing.
- Carefully manage traffic and network design, emphasizing aspects such as standardization and simplification of the technical environments.

Considering that telecom usually represents a significant chunk of the operational cost of a typical call-center operation, the telecom manager has to be cost conscious and has to understand that part of his or her job is to find ways to get more for less from the infrastructure that he or she is responsible for. It is not only matter of making things work but making things work for less.

In summary, it is fundamental to know the actual expenditures of your organization for telecommunications, including telco expenditures, hardware, and services. It is also important to be able to separate the expenditures by identifying not only the services themselves but also things such as penalties for overdue payments, penalties for not achieving minimum-committed traffic, and one-time installation services.

A very common misleading factor when identifying the monthly expenditures of telecom is that very often the telco invoice (bill) encompasses calls from several months (in some countries there is a legal limitation on the time to charge for the calls); therefore, when comparing volumes and costs you should make sure that you separate the calls correctly by time span. In addition, you may also have situations in which the billing cycles of the several providers are not consistent—that means in a given month you may be paying for calls made from different periods. This fact also can generate some distortion when identifying the monthly volume and cost.

Here it is worth mentioning that when verifying the expenditures you should not use as reference the values actually charged but the values that were supposed to be charged. This is necessary because, if you don't, you risk creating the wrong reference for the real expenses. It happens because telcos are well known for their charging mistakes (for and against their clients), and sometimes these mistakes can be very misleading regards having a clear view about how much and on what your organization spends on telecom.

You can generally divide the telecom costs of a call center into three groups:

- Hardware costs
- Personnel costs
- Transport costs

1.1   The hardware costs

The hardware costs encompass all costs associated with the hardware deployed by the call center. Usually, these costs are partially hidden, because some equipment is bought and not rented or leased. This hides the total expenditure and cost of the hardware because part of this cost isn't expended and accounted for monthly.

The telecom manager must keep a careful inventory of the resources used, in which the monthly costs of each device is clearly identified, including the rent/leasing, maintenance, and support costs. When the equipment is owned by the organization, it still has a "cost," which is represented by the total paid (capital invested) or financed monthly by a given rate—usually the WACC (Weighted Average Cost of Capital) of the organization. Knowing exactly how much your hardware costs monthly is very important not only to define your operating baseline properly but also because there is a delicate balance act between investment in hardware and ability to manage traffic.

1.2   The personnel costs

The personnel costs encompass the costs associated with the people directly involved in managing the telecom infrastructure. It includes regular employees, contractors, and expenses for providers of services such as external NOC (Network Operation Center), help desk, network security, and mantainnance.

This cost has to be very well understood, because, as in the cost of the hardware, there is a trade-off between the cost of the control (and the people involved to execute it) and the gains attainable by controlling. The cost of the people may not be an expensive item in some countries; however, it is always important to know it well and to do the math.

1.3 The transport costs

One important aspect of the infrastructure costs, which very often is overlooked, is the transport cost. These costs encompass all the costs involved in transporting the calls between the call center and the user (or vice versa). There are two main reasons these costs very often are not carefully verified:

1) The call-center services very often are outsourced, and in these situations, usually the company contracting the service pays the bills directly, which obscures the visibility of this particular cost item. This situation is common in both inbound and outbound operations.

2) Other very typical situation is when the caller pays for the call (totally or partially). This scenario also tends to obscure the responsibility of the call-center operator in the transport costs.

Although both scenarios are common, keeping track of the transport costs is very important even when you don't actually pay for them directly. Firstly, the transport costs usually amounts to a substantive part of the infrastructure costs (regardless of who foots the bill) and the infrastructure manager usually is the person best able to analyze and optimize this cost.

Therefore, having a clear understanding of your transport costs enables you to act to reduce it.

1.4 Negotiating with service providers

The intensity and frequency of the negotiation processes is the distinct difference between the way telecom is managed in a typical large organization and the way it is in a large call center. Telecom costs in a call center are basic inputs and as such have to be carefully managed.

Therefore, negotiation with service providers is an ongoing process in a large call-center operation. Negotiation skills are assumed to be one of the basic skills of the typical call-center telecom manager.

Even if you have ongoing contracts with a lengthy renewal time, you probably will always have some sort of negotiation going on—new links being added, new tariffs being offered, new clients coming in with their links—which in time will be transferred to you.

Because of this permanent interaction with service providers, telecom managers have to keep track of their organizations' needs very carefully. Therefore, it is absolutely crucial to have a very good, current view of the traffic in terms of volumes, interest, and profile. Spreadsheets like the one shown below have to be updated often and known by heart:

Type of call	Calls	Minutes	Value	% minutes	% Value
Collect calls	2,660	2,399.00	0.00 €	0.01%	0.00%
fix to mobile Intra state calls	135,220	220,551.00	168,161.59 €	0.90%	5.68%
fix to mobile Inter state calls	454,416	472,780.00	462,995.68 €	1.93%	15.65%
fix to mobile Local	314,205	486,268.00	369,043.78 €	1.98%	12.47%
Fix intra state	4,017,093	3,877,638.00	404,831.09 €	15.80%	13.68%
Fix inter state	11,752,302	10,003,687.00	998,390.48 €	40.77%	33.74%
Fix region	477,263	437,190.00	43,586.61 €	1.78%	1.47%
Fix - Local	5,667,798	5,190,523.00	488,815.22 €	21.15%	16.52%
International	3	16.00	51.00 €	0.00%	0.00%
Incoming calls local	708,539	3,690,752.00	0.00 €	15.04%	0.00%
Incoming calls long distance	41,322	156,267.00	23,117.41 €	0.64%	0.78%
tool free	543	974.00	0.00 €	0.00%	0.00%
TOTAL	23,571,364	24,539,052.00	2,958,992.86 €		

The negotiation processes must take into consideration several key factors that are typically driven by the makeup and culture of the organization. As an example, you must evaluate the flexibility and willingness of the organization to absorb new clients and new functions. Some organizations are willing to sell services and then hurry up to expand their infrastructure to support the new clients/services. Others prefer to sell only if there is capacity available and then expand the infrastructure to keep spare capacity and then sell again.

From the above commentary it is clear that some sort of strategy and policy is required to deal with service providers. The need for strategy grows as the size and geographical dispersion of the organization grows. The strategy and policy should address at least the following points:

- Clear statement of global approach to service providers. Typical examples are:

    o One global provider (or one provider by country or region)
    o Providers must have local capabilities and infrastructure in areas of operation
    o Best use of local providers in their areas of operation
    o Local or a universal language capability in regions of the globe
    o Finance capabilities, currency requirements, and payment terms
    o Provisioning and sales-cycle timeframes

- Clear internal understanding and communication regarding functions per type of device, capacity-planning standards, and QoS requirements
- Position the ideal topology based on flow analysis and geographical realities driven by political, organizational, and other regional influences
- Position at least two transport strategies. This might have to be regionally based.
- Define specific contractual elements up front. These could include:

    o Penalties and contract cancellation conditions
    o Minimum-committed volumes
    o Accounts payable and dispute arrangements
    o SLA (Service Level Agreement) components
    o Technology refresh options
    o Help-desk performance metrics

- CPE (Customer Premises Equipment) ownership and management, these might be dictated by regional regulations
- Standard price lists—this is often desirable in decentralized organizations and has many advantages in the areas of cost appropriation and financial planning. This price averaging is often only possible in regional settings, and even though it might

provide a simplified operational structure, it is inevitable that some operational entities will pay a price penalty. That said, the advantages to central planning, financial visibility, and overall simplicity in a large multinational call center are significant.

It is our experience that frequently the people executing the negotiation with the service providers do so without proper preparation. It's is also very common to find people negotiating telco contracts whose view is purely economic without any deep understanding of the traffic volumes, interest, and profile. This lack of preparedness usually generates two very typical approaches:

- "Kick the telco approach"—there is no complication; we quote with all providers and just choose the cheaper ones.
- "Minimalist approach"—there is no need for detailed specification; just ask them for their price list.

These views, although not entirely wrong and may work for a smaller infrastructure, do not take in to consideration a fundamental factor of this process: when negotiating with service providers for a large call center, you are not only comparing the prices per service but also comparing the prices of the different transport strategies (deployment of POPS [Points of Presence] for example) and balancing between the tariffs and the minimum-committed volumes.

Examples:

- Comparing the prices against the cost of building a network of POPs to collect or distribute the traffic; the cost of the spoken minute has to be compared not only against the other service providers' prices but also with the cost to transport them through a private network.
- Mobile traffic: the cost of the spoken-minute fix-mobile cannot be compared only with the fix-trunks' providers but also with the alternative of providing dedicated mobile trunks (or using GSM gateways) to transport mobile traffic.

Therefore, it is crucial to accurately map the traffic, understanding perfectly from where to where the traffic flows, the volumes, the quality requirements, and the available transport strategies.

The fact that we are not just comparing the costs per type of call but also comparing costs per transport strategy is a fundamental point and does make a difference when negotiating with telcos. This is because, besides the possibility of comparing different transport strategies, the only other alternative is the direct comparison between the same services. In this situation each telco usually knows the limits of the others, and the chances for driving down costs are limited (usually associated with the volumes). Therefore there is less room for discounts.

Although large call centers may have several telecom providers, usually they concentrate their business in a few of them (usually between two and three providers are responsible for 80 percent of all telecom expenditures).

It is an important aspect to be considered; we often hear comments like "We don't want to have several providers." The basic fact is that from the point of view of getting better prices it is always convenient to allow all providers to offer their services, even when they are not able to cover all your sites or offer good prices for all services (here the routing capability shows its importance).

Ideally, the logical approach should be: "Provider, quote your best price for the services you offer where you have coverage." Of course, having the ability to route traffic makes all the difference in the effectiveness of this approach.

Here it is important to emphasize that adopting the logic of allowing all potential providers to present their best prices where they have coverage (with no obligation to quote all services for all points of presence) does not necessarily mean contracting in this way. Eventually, you can do a quotation following this strategy, and subsequently, the providers may be short-listed and confronted with the best prices available in each area.

Although common, contracting only one service provider may not be advisable. Having at least two main providers may be better for technical and commercial reasons. First, there is the fact that from a relationship point of view, it is always good to create a situation in which a service provider knows that there is a concurrent active contract to which the organization can easily turn for its requirements.

Another very common argument against segmenting quotations is that such segmentation tends to reduce volumes and consequently tends to reduce the discounts offered. This line of thinking has several problems:

1) The first, and most obvious, is the fact that quoting with several providers doesn't eliminate the possibility of the providers that are able to offer all services in all areas to do so, eventually giving you alternative costs per volume contracted.

2) This line of thinking doesn't take in consideration the already mentioned fact that we have to compare not only service providers and services but also transport strategies; one transport strategy may be applicable only in some areas.

3) And, finally and maybe the more problematic, is the simple fact that we may have potential providers whose prices are very good but limited to some specific geographical area or specific types of services (mobile calls, for example). Depending on the percentage of our services that are these types and within these areas even if we pay a lot more for the other services outside those areas (not covered areas), the savings could still be substantial. In addition, we always can route the calls to avoid using the expensive types of services by redirecting them to providers whose prices are more reasonable.

The quote should be structured in such a way that the service provider must present a defined price that clearly indicates:

- Cost per minute per type of call
- Charging granularity
- Monthly fee for trunk subscription (if any)
- Minimum-committed volume (if any)
- Costs for other services (for example, installation)
- Periodicity and duration of the contract (typically monthly and not above thirty-six months)

Eventually, different minimum-committed volumes imply a different set of tariffs. If this is the case, it has to be clearly stated. The clear indication of all these items enables visibility and transparency when managing and comparing the contract.

These are typically some of the factors to be evaluated regarding the organization itself and its needs, the environments in which it operates its sites, and the providers available when planning for a negotiation process:

- Agility required from the provider companies; telecommunications companies are very process oriented (for good reason), which results in organizations that are not very dynamic in their mode of operation
- Organizational capabilities of the organization in different countries or regions
- Degree of centralization or decentralization of the telecom infrastructure and decision making
- Degree of telecommunications deregulation in countries of operation.
- Language barriers
- Degree of local knowledge of telecommunications industry in other areas of the globe
- Ability or desire to manage multiple vendors and finance systems with different currencies, languages, cultures, and accounts-payable environments
- Ability or desire to route calls in order to always use the cheapest alternative
- Differences of control in regions of the world over different technologies, such as voice, mobile voice, and data networking; these differences are often due to an organization's historical growth and acquisition strategy
- Management's view of outsourcing of personnel

1.5 Interconnection costs (tariffs)

It is basic to know the tariffs when managing a telecom structure in a call-center operation. You must know exactly how much each type of call costs if made through each one of the available service providers with active contracts. You must prepare and keep updated spreadsheets such as these:

Traffic fix-to-fix:

Type	Telco 1 Cost per minute	Telco 1 Granularity sec	Telco 2 Cost per minute	Telco 2 Granularity sec	Telco 3 Cost per minute	Telco 3 Granularity sec	Telco 4 Cost per minute	Telco 4 Granularity sec
Local	USD 0.050	30+6+6	USD 0.027	30+6+6	USD 0.053	60+60	USD 0.04	60+60
Long distance POP	USD 0.051	30+6+6	USD 0.083	6+6+6	USD 0.077	30+6+6	USD 0.10	60+6+6
Long distance outside POP	USD 0.113	30+6+6	USD 0.083	6+6+6	USD 0.077	30+6+6	USD 0.10	60+6+6

Traffic fix-to-mobile:

Tipo	Telco 1 Cost per minute	Telco 1 Granularity sec	Telco 2 Cost per minute	Telco 2 Granularity sec	Telco 3 Cost per minute	Telco 3 Granularity sec	Telco 4 Cost per minute	Telco 4 Granularity sec	Telco 4 Cost per minute	Telco 4 Granularity sec
Mobile local	USD 0.506	30+6+6	USD 0.66	30+6+6	USD 1.12	30+6+6	USD 0.32	30+6+6	USD 0.32	30+6+6
Mobile inter and intra state	USD 0.506	30+6+6	USD 0.66	6+6+6	USD 1.12	30+6+6	USD 0.55	30+6+6	USD 0.43	30+6+6

1.6 Benchmarking

The telecom manager must have a deep awareness of the telecom costs in the markets where he or she has operations. This awareness must be formally organized in spreadsheets in such way as to allow a clear understanding of the negotiable possibilities. Therefore, any hint about prices given by other telecom managers, consultants, and proposals made, should be quickly incorporated in spreadsheets comparing your current price with the new ones. Very often this "market intelligence" is bought—there are companies specialized in scanning the market and mapping the prices of the several services. If your company operates in only one country, it is quite certain that you already have a good view of the prices practiced there. However, even then, it is a good practice to document these prices and periodically share this information with other companies (even if only informally). If you feel like you are working with wrong information, buying "market intelligence" may be a good idea.

This type of comparison doesn't pretend to be absolute and has as a main objective to situate you in terms of minimum and maximum prices usually practiced in each market where your company operates. Of course, you should be aware that the reference prices have to be gotten in a hard-bargain process, but they should be seen as achievable targets.

The spreadsheet below shows a comparison between the current prices of a given organization and the potential prices in a given marketplace (minimum and maximum prices).

Type	Tariff with taxes	Minutes (x1.000)	Current cost (USD)	%	Values of reference minimum (USD)	Values of reference maximum (USD)	Range of values minimum (USD)	Range of values maximum (USD)
			Outbound local					
Fix-Fix	0.0615	213	13,094.40	4.62%	0.0252	0.11	5,375.15	23,435.05
Fix-mobile	0.5532	19.7	10,917.60	3.85%	0.51	0.65	10,065.67	12,828.79
			Outbound long distance					
Fix-Fix	0.0701	459.8	32,219.97	11.37%	0.0252	0.11	11,600.22	50,575.69
Fix-mobile	0.6307	33.8	21,327.35	7.53%	0.51	0.65	17,246.01	21,980.21
			800					
Fix-Fix	0.0561	277.5	15,557.44	5.49%	0.0252	0.11	7,001.47	30,525.65
Fix-mobile	0.0561	66.8	3,743.34	1.32%	0.0252	0.11	1,684.65	7,344.90
Total			96,860.11	34.19%			52,973.17	146,690.28

As can be seen, this spreadsheet allows the quick calculation of the total cost of the organization if the tariffs were given values. In this particular example the current cost spins around USD 96,860. Through the spreadsheet we can see that using the minimum and maximum values practiced in this market the expenditure could be between USD 52,973 and USD 146,690. That means that the current prices are slightly below the average (USD 99,831). Therefore, it is clear that there is some potential for savings through tariffs re-negotiation. If the organization manages to negotiate the minimum tariffs available in this market, it could reduce its current costs by 45 percent (the difference between USD 96,860 and the minimum USD 52,973.17). The telecom manager has to have this kind of understanding, even if he or she can't do anything to change the scenario right away.

1.7   Penalties and other charges

One aspect of managing telecom costs, which is very often overlooked, is what we could call "other charges." These charges include:

- Penalties for overdue payment
- Interests for overdue payment
- Charges for not complying with the minimum-committed volume
- Trunks subscription fees
- Calls made through others service providers (common in countries where you can select the service provider through the prefix dialed)

Here it is important to state the obvious: it is very important to pay the bills on time, and even if no recalculation of the bills is executed every month, some sort of control is necessary, even if only a simplified verification and comparison with the historical values. It is absolutely basic to have a control of what invoices are usually paid, their due dates, and their typical values. If, for any reason, you don't receive the bill, it doesn't absolve you of the penalties associated with late payment.

1.8   Taxes

In some countries (and states, provinces, and cities within these countries) there are special tax breaks for call-center operations. Call centers are usually very labor intensive, and governments want to stimulate the creation of jobs. That usually means you may be entitled to discounts or rebates for your telecom bills. Being aware of the existing tax laws is crucial and can be very well the defining reason in selecting the location of your sites. Very often, telecom managers overlook these possibilities.

Chapter 2: Traffic Management

It is crucial to keep a close eye on the traffic in any call-center operation. It is very important to know the number, type, duration in minutes, and cost of the calls.

In addition, it is absolutely necessary to check the sources and destinations of the traffic to identify from where and to where the calls go. The close control of these data makes possible the identification of strategies that may be deployed to reduce costs and negotiate properly with the service providers. The traffic must be controlled considering the following aspects:

- Traffic volumes (quantity, minutes, and costs)
- Traffic interest (from where to where—area codes, countries, states, cities)
- Typical hourly distribution TCF (Traffic Concentration Factor)
- Typical daily distribution
- Interconnection costs (tariffs)
- Volumes per sites and services (in/outbound)
- Traffic by type (example: outbound to mobile, inbound from mobile)
- Traffic by service provider (originated and terminated)
- Traffic to mobile phones terminated per telco
- Calls duration patterns (typical duration)
- Routing traffic (least-cost routing)
- Traffic concentration (quantity of numbers originating and receiving the calls)

Therefore, all planning initiatives rely on traffic management. Without knowing how your traffic behaves it is hard to negotiate things like tariffs, charging granularities, and minimum-committed volume per contract.

When identifying opportunities for telco cost reduction, the call-center manager has to consider several factors, but in general, these factors can be narrowed down to two main lines of action:

1) Negotiating good prices and charging strategies
2) Forwarding the calls properly

In general, most issues spin around or are linked to knowing the types, profiles, and durations of the calls and from where and to where the calls go and getting the best tariffs as possible. Knowing the typical duration of the calls and their hourly distribution during the day helps to identify which charging granularity is more suitable and also helps to negotiate special prices for specific periods of the day.

Charging granularity definition: Charging granularity refers to the minimum amount of time charged per call. For instance, a charging granularity of 30s+6s+6s indicates that, if a call is established, it is charged as if its duration was 30s even if had less than that. After 30s, the service provider charges an entire quantum of 6s each time span less than or equal to 6s (sometimes called pulse). Therefore, using this charging granularity as an example, a call of 18s duration would be charged as 30s. In the same way a call with duration of 31s would be charged as 36s (30s initial, plus 6s); a call with 38s would be charged 42s (30s initial, plus 6, plus 6); and so on.

Therefore, if you have a large percentage of your calls with duration below 30s, which are known as "short calls," and which would be very typical in active call centers, then it makes sense to have a lower granularity. This is why it will make a big difference if you have a charging granularity of 6s+6s+6s, for example.

Each case has its own particularities; however, the concept has to be well understood and the traffic profile well mapped before any negotiation. It is very important to always keep in mind that what really counts is the average cost of the spoken minute and the balancing act between charging granularity/price per minute and average call duration.

Knowing from where to where the calls come and go also makes a huge difference when negotiating the tariffs. It allows you to try to get better prices in those areas where your traffic is greater. This is pretty obvious; nevertheless, it is very often overlooked.

Another important point when negotiating large voice contracts is what is called "minimum-committed volume." Telco contracts usually demand a minimum monthly payment corresponding to a minimum amount of minutes, regardless if they are really used or not. That means you may have, for example, a clause stating that the minimum-committed volume is USD 100,000 month and your cost per minute is USD 0.01. You have to pay at least USD 100,000 every month, even if the amount of minutes really used doesn't get to 10,000,000. Usually, these negotiations are delicate, given the fact that the cost of the minutes usually goes down as the minimum-committed volume goes up. This is why knowing the average volume of the organization and the volume expected to be forwarded through a specific telco at a given specific tariff is crucial in a negotiation process.

At this point there is an interesting strategic consideration. Even when a specific telco has best prices and conditions, it may be wise to not relay too much of the traffic to only one service provider. The ideal scenario is to divide the traffic between at least two providers, due to technical and commercial reasons:

1) Technical reasons. Any provider may experience technical outages, and in these moments it is crucial to have an alternative or at least the possibility of continuing to operate, even partially. In addition, if you have an outbound operation based on dialers, and these dialers are configured to identify bad numbers through the trunks signalling, the time and effort to adjust the equipment to the public trunks isn't negligible and is what makes any shift between telcos not a simple process in a moment of need.

2) Commercial reasons. It is always good to have a ready alternative if some commercial problem arises. Besides, negotiations are part

of the day-by-day business of any large call-center operation, and it is always healthy to make the providers know that there are ongoing contracts to where you can easily shift your traffic if you need to (with the trunks already installed).

For these reasons, be aware of the siren call when an existing provider comes to you with a potential discount if you raise your minimum-committed volume. You may have to re-direct traffic from other providers and thereby rely too much on only one. Besides, you may default the minimum-committed volume of the other contracts and be forced to pay the minimum values. The discount may not be worth it.

The ideal situation would be, for example: If you had 10 million minutes a month. You may have two contracts, each with 5 million with the minimum-committed volume of 2.5 million. That gives you some leeway to shift traffic between the two providers without incurring penalties. Of course, everything depends on the amount of discount for volume and how critical your operation is.

Here it is important to mention the significance of being able to maneuver the outbound traffic. We are going to discuss this aspect later in more detail; however, it is important to keep in mind that being able to redirect the call flows makes a huge difference in the relationship and negotiation with the service providers. If the service provider knows that you don't have quick alternatives in the short term, it will be tempted to play hardball.

2.1 Traffic volumes

It is very important to understand the type of traffic that your organization has (inbound, outbound, or both) and the objectives associated with the generation of calls and the volumes. You also have to understand the seasonality of the traffic and know how it varies during the year. As the telecom manager, you have to have a clear view of the volumes considering several aspects:

By direction of the flows (inbound and outbound):

Type of traffic	Minutes	%
Inbound	115,107.50	1.29%
Outbound	8,831,515.61	98.71%
Total	8,946,623.11	100.00%

By type (to mobile or to fix, local, long distance, international, per area code):

Type of traffic	Minutes	%
Fix to fix local	2,591,501.09	28.97%
Fix to fix long distance	5,888,906.92	65.82%
Fix to mobile local	252,343.34	2.82%
Fix to mobile long distance	213,871.76	2.39%
Total	8,946,623.11	100.00%

In addition, you have to know how many providers you have and percentage of the volume (per type) each one of them carries:

Provider	Type of service	Minutes	Value
Telco 1	Fix to fix local	1,866,772.50	USD 78,663.92
	Fix to fix long distance	4,371,791.00	USD 388,756.16
	Fix to mobile local	176,224.10	USD 183,385.46
	Fix to mobile long distance	85,479.40	USD 65,560.80
Telco 2	Inbound fix to fix local	34,917.00	USD 7,573.41
	Inbound fix to fix long distance	79,353.40	USD 20,014.90
	Inbound fix to mobile local	458.90	USD 34.20
	Inbound fix to mobile long distance	378.20	USD 29.01
	Outbound fix to fix local	36,124.30	USD 961.61
	Outbound fix to fix long distance	144,466.40	USD 14,431.86
Telco 3	Fix to fix local	101,444.60	USD 15,087.67
Telco 4	Fix to mobile local	23,528.40	USD 9,246.23
	Fix to mobile long distance	102,727.00	USD 71,100.43
Telco 5	Fix to fix local	552,242.69	USD 23,216.65
	Fix to fix long distance	1,293,296.12	USD 114,736.42
	Fix to mobile local	52,131.94	USD 54,123.88
	Fix to mobile long distance	25,287.16	USD 19,349.43
Total		8,946,623.11	USD 1,066,272.04

It is also very important to know the traffic as a whole, but separated into the part paid by your organization directly, the part paid directly by the clients, and the part paid by your organization and later reimbursed by the clients.

2.2 Traffic interest

You have to identify the calls by location in terms of number of calls, duration in minutes, and value. This information allows you to understand the traffic behavior and gives us the tools to act.

City of origin	Area code	City of destination	Area code	Minutes	Value	% Minutes	% Value	Minutes HMM	Erlangs
SAO PAULO	011	SAO PAULO	011	2,824,829.67	USD 226,621.36	31.57%	21.25%	20,994.84	350
SAO PAULO	011	RIO DE JANEIRO	021	799,627.61	USD 110,549.20	8.94%	10.37%	5,943.03	99
SAO PAULO	011	BELO HORIZONTE	312	364,078.10	USD 50,097.65	4.07%	4.70%	2,705.92	45
SAO PAULO	011	CAMPINAS	192	282,486.15	USD 38,906.67	3.16%	3.65%	2,099.51	35
SAO PAULO	011	SALVADOR	712	212,237.11	USD 29,179.26	2.37%	2.74%	1,577.40	26
SAO PAULO	011	BRASILIA	612	191,720.98	USD 26,050.30	2.14%	2.44%	1,424.92	24
SAO PAULO	011	CURITIBA	412	160,862.38	USD 21,742.90	1.80%	2.04%	1,195.57	20
SAO PAULO	011	RECIFE	812	157,832.85	USD 21,456.71	1.76%	2.01%	1,173.05	20
SAO PAULO	011	FORTALEZA	852	133,827.52	USD 18,334.14	1.50%	1.72%	994.64	17
SAO PAULO	011	PORTO ALEGRE	512	125,930.83	USD 16,978.61	1.41%	1.59%	935.95	16
SAO PAULO	011	BLUMENAU	473	111,104.37	USD 15,106.71	1.24%	1.42%	825.76	14
SAO PAULO	011	CAMAQUA	517	110,346.74	USD 14,996.20	1.23%	1.41%	820.12	14
SAO PAULO	011	ARARAQUARA	162	95,557.73	USD 13,047.09	1.07%	1.22%	710.21	12
SAO PAULO	011	BELEM	912	94,837.54	USD 13,129.58	1.06%	1.23%	704.86	12
SAO PAULO	011	LINHARES	272	90,167.62	USD 12,336.45	1.01%	1.16%	670.15	11
SAO PAULO	011	SANTOS	132	81,779.77	USD 11,216.45	0.91%	1.05%	607.81	10
SAO PAULO	011	GOIANIA	622	79,521.08	USD 10,771.01	0.89%	1.01%	591.02	10
SAO PAULO	011	BARRETOS	173	77,013.04	USD 10,648.28	0.86%	1.00%	572.38	10
SAO PAULO	011	Others		2,952,862.01	USD 405,103.47	33.01%	37.99%	21,946.41	366
Total				8,946,623.11	USD 1,066,272.04			66,493.54	1108

The pie graphic gives a better view:

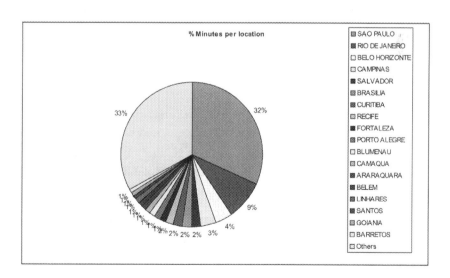

Note in this particular example that only ten area codes concentrate 58.72 percent of the traffic. It, at the very least, sets the direction for an eventual negotiation:

City of destination	Area code	Minutes	Value	% Minutes	% Value
SAO PAULO	011	2,824,829.67	USD 226,621.36	31.57%	21.25%
RIO DE JANEIRO	021	799,627.61	USD 110,549.20	8.94%	10.37%
BELO HORIZONTE	312	364,078.10	USD 50,097.65	4.07%	4.70%
CAMPINAS	192	282,486.15	USD 38,906.67	3.16%	3.65%
SALVADOR	712	212,237.11	USD 29,179.26	2.37%	2.74%
BRASILIA	612	191,720.98	USD 26,050.30	2.14%	2.44%
CURITIBA	412	160,862.38	USD 21,742.90	1.80%	2.04%
RECIFE	812	157,832.85	USD 21,456.71	1.76%	2.01%
FORTALEZA	852	133,827.52	USD 18,334.14	1.50%	1.72%
PORTO ALEGRE	512	125,930.83	USD 16,978.61	1.41%	1.59%
Total		5,253,433.21	USD 559,916.80	58.72%	52.51%

To have a better view of the traffic, you usually divide the traffic by state of destination/origin. It may help if you have telcos that operate predominately in a given state. In our example it is the case of Telefonica de Spana, which operates predominantly in the Brazilian state of São Paulo (SP). As we can see, Telefonica would be particularly well placed to offer a good deal for this particular call-center operator. Note that there are 31.57 percent of the calls within the 011 area (São Paulo city) and 46.71 percent within the state of São Paulo.

State	Quant	Value	Minutes	% Quant	% Value	% Minutes
SP	5,278,500.43	USD 522,079.64	3,939,179.43	46.71%	48.96%	44.03%
RJ	1,074,432.80	USD 107,127.01	976,941.49	9.51%	10.05%	10.92%
MG	880,159.37	USD 80,686.14	741,237.78	7.79%	7.57%	8.29%
BA	478,340.65	USD 43,718.82	402,119.44	4.23%	4.10%	4.49%
RS	534,914.29	USD 42,904.81	399,383.63	4.73%	4.02%	4.46%
PR	461,782.68	USD 38,707.26	358,671.68	4.09%	3.63%	4.01%
SC	381,316.40	USD 31,897.36	295,614.30	3.37%	2.99%	3.30%
PE	296,525.39	USD 26,674.40	245,578.24	2.62%	2.50%	2.74%
CE	283,256.85	USD 26,306.65	241,516.32	2.51%	2.47%	2.70%
GO	211,737.70	USD 17,710.04	164,201.48	1.87%	1.66%	1.84%
DF	215,824.27	USD 17,586.84	163,481.93	1.91%	1.65%	1.83%
ES	178,018.10	USD 16,288.95	149,758.47	1.58%	1.53%	1.67%
PA	149,577.48	USD 15,217.73	138,630.39	1.32%	1.43%	1.55%
RN	127,776.91	USD 10,723.46	99,453.84	1.13%	1.01%	1.11%
MA	92,808.69	USD 9,469.19	86,230.52	0.82%	0.89%	0.96%
PB	85,136.65	USD 8,605.71	78,427.67	0.75%	0.81%	0.88%
AL	108,374.38	USD 8,542.69	79,756.64	0.96%	0.80%	0.89%
AM	84,106.32	USD 8,477.24	77,289.26	0.74%	0.80%	0.86%
MT	79,293.73	USD 6,947.26	64,070.08	0.70%	0.65%	0.72%
MS	78,910.04	USD 5,962.65	55,930.81	0.70%	0.56%	0.63%
PI	60,675.10	USD 5,641.81	51,775.17	0.54%	0.53%	0.58%
SE	52,311.76	USD 5,048.03	46,171.42	0.46%	0.47%	0.52%
RO	38,227.89	USD 3,418.13	31,482.69	0.34%	0.32%	0.35%
TO	26,700.81	USD 2,429.53	22,339.29	0.24%	0.23%	0.25%
AP	17,085.54	USD 1,656.10	15,153.00	0.15%	0.16%	0.17%
AC	12,466.44	USD 1,232.69	11,240.07	0.11%	0.12%	0.13%
RR	11,287.27	USD 1,211.90	10,988.08	0.10%	0.11%	0.12%
Total	11,299,547.97	USD 1,066,272.02	8,946,623.13	100.00%	100.00%	100.00%

The pie graphic gives a better view:

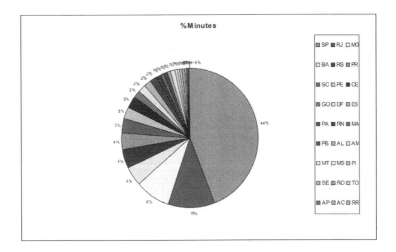

2.3 Traffic hourly distribution HCF (Hourly Concentration Factor)

You must identify the typical hourly distribution, which allows you to identify the hourly concentration factor (HCF), which is the percentage of the daily traffic that occurs during the hour of heaviest traffic. Of course, doing this you also identify which hour has the highest volume.

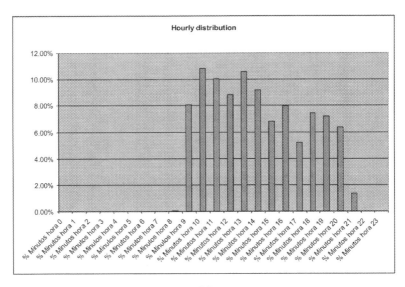

In this particular example, the HCF identified was 10.86 percent and occured between nine and ten o'clock in the morning. The table below gives a better view:

% Minutes Hour 0	% Minutes Hour 1	% Minutes Hour 2	% Minutes Hour 3	% Minutes Hour 4	% Minutes Hour 5
0.00%	0.00%	0.00%	0.02%	0.00%	0.00%
% Minutes Hour 6	**% Minutes Hour 7**	**% Minutes Hour 8**	**% Minutes Hour 9**	**% Minutes Hour 10**	**% Minutes Hour 11**
0.00%	0.00%	0.03%	8.07%	10.86%	10.05%
% Minutes Hour 12	**% Minutes Hour 13**	**% Minutes Hour 14**	**% Minutes Hour 15**	**% Minutes Hour 16**	**% Minutes Hour 17**
8.86%	10.58%	9.17%	6.82%	7.98%	5.20%
% Minutes Hour 18	**% Minutes Hour 19**	**% Minutes Hour 20**	**% Minutes Hour 21**	**% Minutes Hour 22**	**% Minutes Hour 23**
7.45%	7.21%	6.34%	1.34%	0.01%	0.00%

Note that, in this particular case, the traffic outside the typical business hours is almost nonexistent. Therefore, in a negotiation, discounts offered for calls outside business hours have no value whatsoever.

The calculation of this factor is crucial to enable you to do the capacity planning. Based on this factor, you calculate the number of trunks necessary to support the traffic during the busiest hour.

2.4 Traffic daily distribution

To allow you to know how the traffic is distributed along the days of the month, note that we have three curves: 1) number of calls, 2) minutes, and 3) cost.

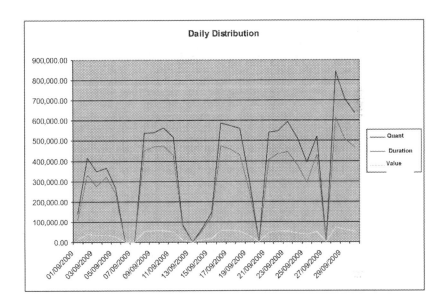

We may see the data through the spreadsheet:

Data	Quant	Minutes	Cost	% Quant	% Minutes MCF day	% Cost
01/09/2009	137,881.94	111,162.78	USD 13,208.01	1.22%	1.24%	1.24%
02/09/2009	416,640.32	331,844.86	USD 39,490.67	3.69%	3.71%	3.70%
03/09/2009	349,392.22	277,207.40	USD 34,265.18	3.09%	3.10%	3.21%
04/09/2009	366,275.72	320,608.93	USD 37,925.73	3.24%	3.58%	3.56%
05/09/2009	264,357.74	239,083.23	USD 27,829.30	2.34%	2.67%	2.61%
06/09/2009	30.10	88.63	USD 58.16	0.00%	0.00%	0.01%
07/09/2009	120.38	132.95	USD 95.27	0.00%	0.00%	0.01%
08/09/2009	538,436.29	451,327.60	USD 53,243.79	4.77%	5.04%	4.99%
09/09/2009	541,852.12	471,249.41	USD 55,099.04	4.80%	5.27%	5.17%
10/09/2009	564,453.74	474,538.90	USD 55,331.19	5.00%	5.30%	5.19%
11/09/2009	516,572.00	427,394.59	USD 50,016.69	4.57%	4.78%	4.69%
12/09/2009	85,034.47	78,955.39	USD 9,711.17	0.75%	0.88%	0.91%
13/09/2009	0.00	0.00	USD 0.00	0.00%	0.00%	0.00%
14/09/2009	64,088.09	52,107.67	USD 8,718.80	0.57%	0.58%	0.82%
15/09/2009	144,969.40	124,885.46	USD 16,441.05	1.28%	1.40%	1.54%
16/09/2009	589,417.85	474,336.32	USD 56,954.46	5.22%	5.30%	5.34%
17/09/2009	572,925.59	457,496.35	USD 56,664.31	5.07%	5.11%	5.31%
18/09/2009	560,240.39	428,711.40	USD 52,881.35	4.96%	4.79%	4.96%
19/09/2009	291,759.58	244,132.69	USD 29,838.14	2.58%	2.73%	2.80%
20/09/2009	0.00	0.00	USD 0.00	0.00%	0.00%	0.00%
21/09/2009	540,633.25	405,181.03	USD 47,831.93	4.78%	4.53%	4.49%
22/09/2009	547,856.14	435,467.65	USD 52,218.48	4.85%	4.87%	4.90%
23/09/2009	594,157.87	446,092.02	USD 53,570.24	5.26%	4.99%	5.02%
24/09/2009	511,019.41	380,241.41	USD 44,557.12	4.52%	4.25%	4.18%
25/09/2009	391,811.64	293,311.73	USD 34,942.02	3.47%	3.28%	3.28%
26/09/2009	521,703.27	431,376.67	USD 50,267.64	4.62%	4.82%	4.71%
27/09/2009	30.10	168.40	USD 121.46	0.00%	0.00%	0.01%
28/09/2009	842,218.97	612,941.92	USD 71,511.01	7.45%	6.85%	6.71%
29/09/2009	708,971.71	509,707.85	USD 59,588.09	6.27%	5.70%	5.59%
30/09/2009	636,697.68	466,869.75	USD 53,891.71	5.63%	5.22%	5.05%
Total	11,299,548.00	8,946,623.00	USD 1,066,272.00			

In this particular example you have a monthly concentration factor (MCF) of 6.85 percent. This number indicates the percentage of the monthly traffic occurs during the day with the biggest volume.

The capacity-planning calculations usually are based on the MCF and HCF. In other words, you calculate how many trunks you need during the busiest hour of the busiest day of the busiest month (if there are different volumes along the months).

2.5   Calls duration patterns

It is important to constantly monitor the calls duration pattern. This verification is important to put in perspective the time of the contact and the associated cost. Usually, short calls (under 30s) include the calls to numbers with faxes or answering machines, in which the dialer discharges the call without transferring it to the attendance group (in automatic-dialing installations). Calls between 30s and 60s may encompass those transferred to the attendance group and discharged before being attended (dropped in line). The attentive monitoring of these parameters, allows verification if there is any problem with the predictive dialer regarding call classification, dialing rate, or even with the mailings themselves. Therefore, it is very important to have a clear view of the duration distribution of the calls. The spreadsheet below is the kind of information that the telecom manager has to have monthly.

From (Min)	to (Min)	Cost	Quant	Minutes	% cost	% quant	% minutes	% minutes accumulated
0.2	0.5	USD 333,259.53	7,528,854.90	3,127,686.46	31.25%	66.63%	34.96%	34.96%
0.5	0.7	USD 68,712.22	1,158,702.75	616,949.26	6.44%	10.25%	6.90%	41.86%
0.7	1	USD 72,477.99	804,378.07	590,209.32	6.80%	7.12%	6.60%	48.45%
1	1.5	USD 82,298.02	617,557.77	651,474.42	7.72%	5.47%	7.28%	55.73%
1.5	2	USD 72,589.56	388,355.13	574,566.59	6.81%	3.44%	6.42%	62.16%
2	3	USD 86,778.64	329,121.18	675,718.11	8.14%	2.91%	7.55%	69.71%
3	4	USD 56,144.43	148,550.65	431,692.09	5.27%	1.31%	4.83%	74.53%
4	5	USD 42,152.04	85,741.07	321,135.08	3.95%	0.76%	3.59%	78.12%
5	6	USD 33,251.08	54,782.92	251,174.44	3.12%	0.48%	2.81%	80.93%
6	7	USD 27,785.96	38,829.13	210,461.88	2.61%	0.34%	2.35%	83.28%
7	8	USD 24,344.35	29,592.73	185,230.25	2.28%	0.26%	2.07%	85.35%
8	9	USD 21,769.71	23,758.41	168,371.62	2.04%	0.21%	1.88%	87.24%
10	60	USD 144,223.73	91,261.48	1,138,089.82	13.53%	0.81%	12.72%	99.96%
60	100	USD 484.55	63.80	3,863.77	0.05%	0.00%	0.04%	100.00%
Total		USD 1,066,272.04	11,299,548.00	8,946,623.11				

The table is better understood through the following graphic:

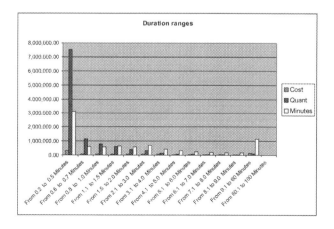

In this example, 76.86 percent of the calls have duration equal to or less than 0.7 minutes and account for 37.70 percent of the cost. The average call duration is 0.79 minutes (47.50s).

If we analyze the average cost of the contact, we may have a spreadsheet such as:

Dutation range	Cost per call	Cost per minute
From 0.2 to 0.5 Minutes	USD 0.04	USD 0.11
From 0.6 to 0.7 Minutes	USD 0.06	USD 0.11
From 0.8 to 1.0 Minutes	USD 0.09	USD 0.12
From 1.1 to 1.5 Minutes	USD 0.13	USD 0.13
From 1.5 to 2.0 Minutes	USD 0.19	USD 0.13
From 2.1 to 3.0 Minutes	USD 0.26	USD 0.13
From 3.1 to 4.0 Minutes	USD 0.38	USD 0.13
From 4.1 to 5.0 Minutes	USD 0.49	USD 0.13
From 5.1 to 6.0 Minutes	USD 0.61	USD 0.13
From 6.1 to 7.0 Minutes	USD 0.72	USD 0.13
From 7.1 to 8.0 Minutes	USD 0.82	USD 0.13
From 8.1 to 9.0 Minutes	USD 0.92	USD 0.13
From 9.1 to 60 Minutes	USD 1.58	USD 0.13
From 60.1 to 100 Minutes	USD 7.59	USD 0.13

Here it is important to mention that you should be able to get the real duration of the calls from your billing system and not only from the bill. This is necessary, because if your charging granularity was, for example, 30s, any call regardless of the duration between 0 and 30s will appear on the bill as having the duration of 30s. The billing system on the other hand will indicate the real duration of the calls. We can group the billing logs by duration in ranges of 6 seconds and have the following view:

Range (secs)	Minutes	% minutes	Calls	% calls
00001 to 00006	46,389.81	3.43%	927,796.11	12.32%
00007 to 00012	551,633.78	40.10%	3,677,558.50	48.85%
00013 to 00018	312,629.99	18.40%	1,250,519.96	16.61%
00019 to 00024	344,884.91	20.14%	985,385.46	13.09%
00025 to 00030	309,417.69	17.93%	687,594.86	9.13%
Total (Bellow or equal 30secs)	1,564,956.17	100.00%	7,528,854.90	100.00%

Through this table it becomes possible to verify that in this particular organization a typical short call was 12.47 seconds. Through this analysis it becomes possible to verify which granularity suits the organization best:

Minutes charged 30+6+6	Minutes charged 6+6+6	Really used
3,127,686.46	2,258,656.47	1,564,956.17

In this particular case we have a situation in which, through a granularity of 30s+6s+6s, the organization was charged a total of 3,127,686.46 minutes, but we know that the average duration of the short calls is in reality 12.47 seconds and therefore if the oganization had a charging granularity of 6s+6s+6s it would pay only for 2,258,656.47 minutes. Of course, that still would be above the real usage of 1,564,956.17; however, it would be a lot better than using 30s+6s+6s granularity. In terms of value, this difference would be:

3,127,686.46 × USD 0.12 = USD 333,259.52

2,258,656.47 × USD 0.12 = USD 271,038.78

That means a savings of USD 62,000 or 18 percent of the total spent on short calls—not a value to be ignored.

2.5.1 Short calls

As we have previously defined them, short calls are those whose duration wasn't enough to make feasible any kind of treatment by the attendant. Regardless of the precise definition, in an outbound operation "short calls" is the name given to the calls in which the dialer device identified that the number wasn't valid or the calls were answered by a machine (answering machine or fax) or the calls, attended by a person, were dropped by the client as soon the salutation ended. In an inbound call center, "short call" is the name given to calls that are dropped before a certain time (usually after the salutation and soon after being placed into the attendance queue).

Whatever the precise definition, the point is this: these calls don't generate a conversation.

Usually, we draw a line on this by time, defining arbitrarily that short calls are the calls dropped by the outbound dialer before thirty seconds time (because of bad numbers, answering machines, and faxes) and the calls where the prospective customer hangs up before the conversation gets into full swing—between thirty seconds and one minute, for example.

In an inbound call center, we may draw a line considering the calls dropped before the end of the salutation and the calls dropped ten seconds after entering the attendance queue and ten seconds after being transferred to an agent.

Note that here you would have three types of short calls with different durations. It is very important to keep track of these calls for several reasons:

1) It is important to know the proportion between the number of calls successfully concluded (total of the calls minus the short calls) and the number of calls tried (total of the calls). This information points to the efficiency of the process of reaching clients in general and, in the case of outbound operations, the quality of the mailings used.
2) It is important to know the cost of trying to reach the clients and the cost of actually talking with them. This information allows the understanding of the cost of having bad mailings.
3) It is important to know if the dialers are operating properly. A big imbalance between the number of tries and the number of successes can indicate that the predictive algorithm isn't working properly and the dialer is producing more calls than the attendants can handle.
4) It is important to be able to compare the agent's effective work time with the number of tries.

To really understand what is happening in your infrastructure you are going have to dig deeper. Depending on the types of calls you are taking (or making), high volumes of short calls can be quite acceptable, especially in an outbound operation (wrong-party contact, no interest, etc.).

If yours is primarily an inbound operation, it is worth understanding your call types and looking at the call stats at both a global and individual level. Look for patterns or anomalies, such as high levels of transferred calls, individual agents with unusually high numbers of short calls, and particular inbound numbers with high numbers of short calls, and drill down from there.

If you record the calls, identify a group of short calls and see if there are any similarities. If you have access to speech analytics, you can do this in volume, but even working manually you should get some insight. Some PBXs and call-recording solutions (or call loggers) allow you to see which party terminated the call. There are no industry averages—most call centers tend to have different experiences, even within the same industry. You are going to think about what types of calls you take or make and consider what the likely customer expectations would be. It will give you some insight about what level of short calls is acceptable.

It is very important to have monthly spreadsheets such as the one shown below:

time span (secs)	Calls	Minutes	Cost	% Calls	% accumulated calls	% minutes	% accumulated minutes	% cost	% accumulated cost
00000 a 00003	523,329.00	13,300.00	USD 79,545.91	2.15%	2.15%	0.06%	0.06%	1.82%	1.82%
00004 a 00005	782,036.00	66,857.00	USD 99,098.43	3.21%	5.37%	0.28%	0.34%	2.55%	4.36%
00006 a 00009	1,067,636.00	143,734.00	USD 90,747.57	4.39%	9.75%	0.61%	0.95%	2.33%	6.70%
00009 a 00012	3,741,951.00	729,761.00	USD 394,766.70	15.38%	25.13%	3.10%	4.06%	10.16%	16.86%
00013 a 00018	3,060,121.00	784,356.00	USD 207,004.50	12.58%	37.71%	3.34%	7.40%	5.33%	22.18%
00019 a 00024	2,314,978.00	827,882.00	USD 157,160.41	9.51%	47.23%	3.52%	10.92%	4.04%	26.23%
00025 a 00030									
00031 a 00060	6,358,839.00	4,513,260.00	USD 532,935.13	26.14%	81.39%	19.22%	33.93%	13.71%	43.21%
00061 a 00090	1,732,081.00	2,111,047.00	USD 279,953.31	7.12%	88.50%	8.99%	42.91%	7.20%	50.41%
00091 a 00120	861,690.00	1,498,263.00	USD 229,960.38	3.54%	92.05%	6.38%	49.29%	5.92%	56.33%
00121 a 00180	743,403.00	1,796,789.00	USD 327,999.88	3.06%	95.10%	7.65%	56.94%	8.44%	64.77%
00181 a 00240	325,213.00	1,124,499.00	USD 207,412.09	1.34%	96.44%	4.79%	61.73%	5.34%	70.10%
00241 a 00300	192,522.00	860,813.00	USD 149,395.70	0.79%	97.23%	3.66%	65.39%	3.84%	73.95%
00301 a 00600	387,239.00	2,690,770.00	USD 382,243.41	1.59%	98.82%	11.46%	76.85%	9.83%	83.78%
00601 a 00900	134,547.00	1,640,094.00	USD 219,260.94	0.55%	99.37%	6.98%	83.83%	5.64%	89.42%
00901 a 01200	67,353.00	1,161,435.00	USD 163,113.14	0.28%	99.65%	4.94%	88.78%	4.20%	93.62%
01201 a 09999	84,416.00	2,526,904.00	USD 242,861.56	0.35%	100.00%	10.76%	99.54%	6.25%	99.87%
10000 a 86400	512	108,649.00	USD 5,156.60	0.00%	100.00%	0.46%	100.00%	0.13%	100.00%
TOTAL	24,329,780	23,487,969.00	USD 3,886,602.73	100.00%		100.00%		100.00%	

2.6   Concentration of traffic

The traffic concentration per destination/origin is another key performance indicator to be monitored in call centers. We must be able to see how many numbers to and from the calls were made, and we must be able to identify how many numbers were called or called only one time, how many more than one time, and how calls were distributed by quantity of calls (quantity grouped).

Example:

Outbound						
Item		Quant	Month			
			Minutes	%	Cost	%
Calls made		19,703,828				
Numbers dialed		11,144,732	%	17,522,963.00	%	USD 2,491,412.03	%
Numbers dialed more than one time		4,049,393	36.33%	10,380,538.33	59.24%	USD 1,358,615.68	54.53%
Numbers dialed only one time		7,095,338	63.67%	7,142,424.67	40.76%	USD 1,132,796.35	45.47%
Cost associated with the repetition		12,608,490	63.99%	6,331,145.13	59.24%	USD 922,277.20	37.02%
Average number of calls per number (including the ones with one call)		1.77				
Average number of calls per number (only ones with repetition)		3.11				

This particular spreadsheet shows that 63.67 percent of the numbers called (it is an outbound operation) were called only one time. The average number of calls per number was 1.17, but, if we remove the numbers called only one time, we have an average calls per number of 3.11.

As we can also see 37.02 percent of the total cost is due to repetitive calls. The difference between 63.99 percent (quantity of repeated calls) and 37.02 percent (cost of the repeated calls) points to the fact that these calls probably are short calls. This spreadsheet tells us that 37.02 percent of all costs occur without talking with the clients, but merely trying to reach them.

As mentioned before, it is also very important to group the quantity of the numbers per number of calls:

Number of calls	Number of numbers dialled	Duration (minutes)	Cost	% quant	% Minutes	% Cost	Accumulated % Quant	Accumulated % Duração	Accumulated % Valor
1	7,095,338	7,142,424.32	USD 1,132,796.34	36.01%	40.76%	45.47%	36.01%	40.76%	45.47%
2	4,425,169	3,939,849.64	USD 531,962.30	22.46%	22.48%	21.35%	58.47%	63.24%	66.82%
3	2,484,202	2,054,360.13	USD 265,355.27	12.61%	11.72%	10.65%	71.08%	74.97%	77.47%
4	1,711,437	1,381,461.85	USD 177,508.65	8.69%	7.88%	7.12%	79.76%	82.85%	84.60%
5	1,071,103	850,256.18	USD 108,290.64	5.44%	4.85%	4.35%	85.20%	87.70%	88.94%
6	813,544	621,544.91	USD 79,877.35	4.13%	3.55%	3.21%	89.33%	91.25%	92.15%
7	528,488	398,229.04	USD 51,190.21	2.68%	2.27%	2.05%	92.01%	93.52%	94.20%
8	433,347	322,682.11	USD 41,202.40	2.20%	1.84%	1.65%	94.21%	95.37%	95.86%
9	272,597	199,272.19	USD 26,161.84	1.38%	1.14%	1.05%	95.59%	96.50%	96.91%
10	236,956	165,936.42	USD 20,873.14	1.20%	0.95%	0.84%	96.79%	97.45%	97.74%
11	157,522	110,147.52	USD 13,915.22	0.80%	0.63%	0.56%	97.59%	98.08%	98.30%
12	137,171	94,476.76	USD 11,744.94	0.70%	0.54%	0.47%	98.29%	98.62%	98.77%
13	89,111	63,799.71	USD 7,980.62	0.45%	0.36%	0.32%	98.74%	98.98%	99.09%
14	68,547	48,061.38	USD 5,968.05	0.35%	0.27%	0.24%	99.09%	99.26%	99.33%
15	44,211	30,844.25	USD 3,843.49	0.22%	0.18%	0.15%	99.31%	99.43%	99.49%
16	34,593	22,862.03	USD 2,848.25	0.18%	0.13%	0.11%	99.49%	99.56%	99.60%
17	23,240	16,420.99	USD 1,974.98	0.12%	0.09%	0.08%	99.61%	99.66%	99.68%
18	16,230	10,688.58	USD 1,322.33	0.08%	0.06%	0.05%	99.69%	99.72%	99.74%
19	12,526	8,793.09	USD 1,116.29	0.06%	0.05%	0.04%	99.75%	99.77%	99.78%
20	10,083	6,110.98	USD 758.36	0.05%	0.03%	0.03%	99.81%	99.80%	99.81%
21	7,126	4,944.92	USD 609.23	0.04%	0.03%	0.02%	99.84%	99.83%	99.83%
22	5,759	3,600.40	USD 442.77	0.03%	0.02%	0.02%	99.87%	99.85%	99.85%
23	3,345	2,696.84	USD 330.62	0.02%	0.02%	0.01%	99.89%	99.87%	99.87%
24	3,490	2,685.57	USD 328.17	0.02%	0.02%	0.01%	99.91%	99.88%	99.88%
25	4,121	2,590.28	USD 314.36	0.02%	0.01%	0.01%	99.93%	99.90%	99.89%
26	1,008	475.61	USD 58.76	0.01%	0.00%	0.00%	99.93%	99.90%	99.89%
27	1,571	800.47	USD 97.27	0.01%	0.00%	0.00%	99.94%	99.90%	99.90%
28	814	465.21	USD 61.98	0.00%	0.00%	0.00%	99.94%	99.91%	99.90%
29	562	467.81	USD 57.23	0.00%	0.00%	0.00%	99.95%	99.91%	99.90%
30	291	129.95	USD 24.09	0.00%	0.00%	0.00%	99.95%	99.91%	99.90%
31	601	382.91	USD 45.41	0.00%	0.00%	0.00%	99.95%	99.91%	99.91%
32	310	149.87	USD 16.26	0.00%	0.00%	0.00%	99.95%	99.91%	99.91%
33	640	315.34	USD 39.28	0.00%	0.00%	0.00%	99.96%	99.91%	99.91%
34	659	323.14	USD 57.23	0.00%	0.00%	0.00%	99.96%	99.92%	99.91%
36	349	160.27	USD 17.18	0.00%	0.00%	0.00%	99.96%	99.92%	99.91%
40	776	388.11	USD 65.66	0.00%	0.00%	0.00%	99.96%	99.92%	99.91%
48	465	409.77	USD 49.40	0.00%	0.00%	0.00%	99.97%	99.92%	99.92%
66	640	575.23	USD 74.56	0.00%	0.00%	0.00%	99.97%	99.92%	99.92%
70	679	307.54	USD 305.15	0.00%	0.00%	0.01%	99.97%	99.93%	99.93%
95	921	418.43	USD 44.65	0.00%	0.00%	0.00%	99.98%	99.93%	99.93%
110	1,066	506.79	USD 85.46	0.01%	0.00%	0.00%	99.98%	99.93%	99.94%
332	3,219	11,945.61	USD 1,596.64	0.02%	0.07%	0.06%	100.00%	100.00%	100.00%
Total	19,703,828	17,522,962.13	USD 2,491,412.00						

This spreadsheet shows that 22.53 percent of the cost is concentrated in numbers that demand more than three calls. In this particular example, we can see a noticeable fact that around 1,574,547 numbers (more than seven tries) out of 19,703,808 (8 percent of the total) correspond to 5.8 percent of the cost.

This kind of information is very helpful from many different aspects. For instance, the identification of a high quantity of short calls to the same number may indicate configuration problems in the dialers: a bad adjustment between dial rate and availability of agents or poor bad-number identification.

Of course, a high number of repetitions to the same number may be just poor management of the mailings by the operational people, having nothing to do with the telecom devices. Nevertheless, it still is a potential problem, and the telecom manager is the professional best placed to spot the phenomenon, even if the resolution is beyond his or her scope.

2.7   The type of traffic (fix-to-fix/fix-to-mobile)

In many countries, the tariff system differentiates clearly between the costs associated with calls made among the fix-lines network (fix-to-fix), the calls among mobile lines (mobile-to-mobile), and calls made between these two networks (fix-to-mobile or mobile-to-fix).

This fact obligates the telecom managers in large outbound operations in these countries to identify the volumes and costs associated with each type of traffic. It is important to know the volume and cost associated with calls made to fix lines and the volumes and costs associated to calls made to mobile phones. This information should be identifiable as shown below:

Type of service	Minutes	%	Cost	%
Fix-to-fix	8,480,408.01	94.79%	USD 663,442.60	62.22%
Fix-to-mobile	466,215.10	5.21%	USD 402,829.44	37.78%
Total	8,946,623.11	100.00%	USD 1,066,272.04	100.00%

As we can see, for this particular example calls to mobile phones represent 5.21 percent of the total but account for 37.78 percent of the costs. Usually, calls between fix lines and mobile lines (or vice versa) cost a lot more than calls between lines of the same type. There are several strategies that can be deployed to circumvent this tariff system peculiarity (GSM gateways dedicated mobile trunkson-net and on-group selection), which will be discussed subsequently in this book; however, identifying the traffic pattern is the first step to be able to use any of them.

2.8 Mapping the mobile traffic per service provider (outbound)

When negotiating with a mobile provider, it is easier to get better prices for calls between lines that belong to the provider itself. These calls are usually called "on-net," meaning they don't leave the provider network. Usually, the providers can be much more flexible regarding the price of these calls, because they don't have to pay what is known an "interconnection fee," which is basically what a provider has to pay to other networks to transport a call originated in its own network and terminated in another.

In addition, the quality of the calls transported over the same network (on-net) tends to be significantly better than the ones flowing through different provider networks.

These are the reasons why it is so important to map the mobile traffic (mobile traffic meaning the traffic terminated in mobile lines) and inform the providers of the volumes which could be on-net.

Here you have to remember the fact that you may or may not use dedicated mobile trunks or GSM gateways. If you do, the need for mapping the traffic regarding the terminated provider becomes even more important, considering the fact that you may separate the outgoing mobile traffic by routing it through different trunk groups. Proceeding in such a way, you will guarantee that all mobile calls will be always originated and terminated over the same provider (all mobile traffic becomes on-net). You basically identify the provider to which the call is destined and route the call to a trunk group of this particular provider.

The mapping of the traffic by termination allows the calculation of the ROI (Return on Investment) of investing in routing devices and GSM gateways (you need to separate the outgoing traffic per mobile service provider). This strategy can be very effective in terms of reducing mobile-traffic costs. It is not unusual to achieve reductions above 50 percent in some countries. The spreadsheet below shows an example of how this information should be presented:

Provider	Quant	Minutes	Cost	% Quant	% Minutes	% Cost
Telco 1	10,117.68	7,617.77	USD 5,047.10	1.62%	1.63%	1.25%
Telco 2	2,131.90	1,556.31	USD 1,491.75	0.34%	0.33%	0.37%
Telco 3	13,804.33	8,171.50	USD 7,826.88	2.21%	1.75%	1.94%
Telco 4	118,102.37	91,801.23	USD 78,499.55	18.90%	19.69%	19.49%
Telco 5	7,663.84	3,758.75	USD 3,595.70	1.23%	0.81%	0.89%
Telco 6	65,586.26	53,734.85	USD 47,914.91	10.50%	11.53%	11.89%
Telco 7	294.46	156.73	USD 149.97	0.05%	0.03%	0.04%
Telco 8	8,315.58	7,321.87	USD 7,000.83	1.33%	1.57%	1.74%
Telco 9	76,736.51	70,313.36	USD 58,461.07	12.28%	15.08%	14.51%
Telco 10	47.11	23.47	USD 13.34	0.01%	0.01%	0.00%
Telco 11	321,928.20	221,759.26	USD 192,828.34	51.53%	47.57%	47.87%
Total	624,728.23	466,215.10	USD 402,829.44			

The following spreadsheet shows the same information but separated by type:

VC1—Local
VC2—Intrastate
VC3—Interstate

Type	Quant	Minutes	Cost	% Quant	% Minutes	% cost
VC1	150,540.19	104,409.44	USD 56,757.51	24.10%	22.40%	14.09%
VC2	34,620.75	25,382.67	USD 24,265.06	5.54%	5.44%	6.02%
VC3	439,567.29	336,422.99	USD 321,806.85	70.36%	72.16%	79.89%
Total	624,728.23	466,215.10	USD 402,829.42			

In this particular example, it is clear that the relevant tariff is the interstate (VC3). It is also clear that the provider Telco11 terminates most of the calls. This information is crucial in any negotiation.

2.9   Routing traffic

To be able to route your traffic (this usually refers to outbound traffic but is also applicable to inbound) is absolutely crucial. It allows two fundamental things:

1) The forwarding of the calls to the provider, who is able to transport it for the cheaper price
2) Redirecting the call in case of major outage in a particular service provider

The ability to redirect the calls to the provider who is able to transport them for the smallest cost and an understanding of the traffic volume and interest ("interest" meaning to where and from where the calls go) are key points when dealing with service providers.

If you know your traffic profile and interest but you are not able to route the traffic, your negotiation process has to focus on getting lower tariffs for the services with bigger volumes, trying at the same time to avoid having prices too high for the remaining services.

When negotiating with service providers, the possibility of routing calls opens the possibility of extracting better tariffs for specific types of calls and letting others be expensive. This is crucial, given the fact that some service providers rely on others to deliver calls outside their own network. That means they have the possibility of charging less for the calls within their own network. Being able to route allows you to take advantage of situations like that and usually makes it possible to attain a much better overall telco cost.

Of course, being able to route traffic is what makes possible the deployment of private networks with POPs. Without being able to forward the calls to specific trunks or tie-lines (least-cost routing—LCR) the establishment of POPs (from where the calls are collected as inbound traffic or to where calls are forwarded to get the public network near to the destination—outbound traffic) is not possible.

Another very important aspect associated with being able to route calls is the possibility of redirecting them in case of a major outage. That can be very important, depending how critical your operation is. It is particularly important when you have an outbound operation using dialers. Because dialers demand some signaling adjustments, it is hard to shift trunks quickly during an outage. Therefore, to have an effective backup you have to have the trunks connected to the equipment and operating (with the signalling previously adjusted).

Once again, it is highly recommended to have at least two providers. Having only one provider, even if the provider has two different physical-access infrastructures, will still expose your organization to administrative problems that may affect the provider, such as strikes, bankruptcy, and contracts disagreements.

2.9.1 Example of least-cost routing (LCR)

In this example, we will demonstrate the calculations that show how much an organization could be expending if it had the resources to route its traffic properly. Note here the fact that this calculation is done without considering any discount. The calculation analyzes your final telco cost if, with your current contracts, you routed your traffic using your current prices. This particular organization had three contracts with three different service providers:

- Telco 1
- Telco 2
- Telco 3

To execute the calculations properly it was necessary to identify the typical duration of the calls:

Type of call	%
Success calls	65.04%
Short calls	34.96%

Assuming short calls are those whose duration is equal or less than thirty seconds and taking as reference only the fix-to-fix calls:

	Type	Telco 1	Telco 2	Telco 3
Success	Local	USD 0.040	USD 0.027	USD 0.044
calls	Long Distance within POP	USD 0.040	USD 0.090	USD 0.083
	Longa distance out of POP	USD 0.100	USD 0.090	USD 0.083
	Type	Telco 1	Telco 2	Telco 3
Short calls	Local	USD 0.020	USD 0.014	USD 0.022
	Long Distance within POP	USD 0.020	USD 0.027	USD 0.042
	Longa distance out of POP	USD 0.050	USD 0.027	USD 0.042

The item "Long Distance within POP" corresponds to calls to locations where the service provider Telco 1 has POPs, and the calls are charged as local. The traffic volumes are:

		Average duration (secs)	Type	% traffic	Minutes
	Success		Local	18.50%	1,019,769.06
	Calls	92	Long distance within POP	34.00%	1,874,170.17
Fix-to-fix			Long distance out of POP	47.50%	2,618,325.97
		Average duration (secs)	Type	% traffic	Calls
8.480.408 Min.	Short Calls		Local	18.50%	2,196,426
		15	Long distance within POP	34.00%	4,036,674
			Long distance out of POP	47.50%	5,639,471

Considering the current tariffs, the costs would be:

	Type	Telco 1	Telco 2	Telco 3
Success	Local	USD 40,790.76	USD 27,533.76	USD 44,517.00
Calls	Long distance within POP	USD 74,966.81	USD 168,506.64	USD 155,914.09
	Long distance out of POP	USD 261,832.60	USD 235,413.69	USD 217,821.16
	Type	Telco 1	Telco 2	Telco 3
Short Calls	Local	USD 43,928.51	USD 30,749.96	USD 47,941.38
	Long distance within POP	USD 80,733.48	USD 108,881.21	USD 167,907.48
	Long distance out of POP	USD 281,973.57	USD 152,113.46	USD 234,576.63
	Type	Telco 1	Telco 2	Telco 3
Combined	Local	USD 84,719.28	USD 58,283.72	USD 92,458.38
	Long distance within POP	USD 155,700.29	USD 277,387.85	USD 323,821.57
	Long distance out of POP	USD 543,806.16	USD 387,527.15	USD 452,397.79
	Total	USD 784,225.73	USD 723,198.73	USD 868,677.74

If we execute the calls using the cheapest alternatives, we would have the following cost for fix-to-fix calls:

Type	Provider	Cost
Local	Telco 2	USD 58,283.72
Long distance within POP	Telco 1	USD 155,700.29
Long distance out of POP	Telco 2	USD 387,527.15
Total		USD 601,511.16

Now we are going to do the same calculations for calls fix-to-mobile:

Type	Telco 1	Telco 2	Telco 3	Telco 4
Success	VC1	USD 0.550	USD 0.650	USD 0.988	USD 0.395
Calls	VC2 e VC3	USD 0.550	USD 0.730	USD 0.988	USD 0.695
Type	Telco 1	Telco 2	Telco 3	Telco 4
Short Calls	VC1	USD 0.270	USD 0.325	USD 0.494	USD 0.198
	VC2 e VC3	USD 0.270	USD 0.210	USD 0.494	USD 0.348

Using the current volumes:

		Minutes	Average duration (secs)	Type	% tráffic	Minutes
Fix-to-mobile	Success			VC1	22.40%	67,880.92
	Calls	303,039.82	92	VC2 and VC3	77.60%	235,158.90
		Calls	Average duration (secs)	Type	% tráffic	Calls
	Short Calls			VC1	22.40%	73,102.53
		326,350.57	30	VC2 and VC3	77.60%	253,248.04

You would have the costs as follows:

	Type	Telco 1	Telco 2	Telco 3	Telco 4	Telco 5	Gateway GSM
Success	VC1	USD 37,334.61	USD 44,122.60	USD 67,082.32	USD 26,812.96	USD 19,006.66	USD 12,897.37
Calls	VC2 e VC3	USD 129,337.39	USD 171,665.99	USD 232,392.32	USD 163,435.43	USD 89,360.38	USD 49,383.37
	Type	Telco 1	Telco 2	Telco 3	Telco 4	Telco 5	Gateway GSM
Short Calls	VC1	USD 19,737.68	USD 23,758.32	USD 36,121.25	USD 14,437.75	USD 10,234.35	USD 6,944.74
	VC2 e VC3	USD 68,376.97	USD 53,182.09	USD 125,134.33	USD 88,003.69	USD 48,117.13	USD 26,591.04
	Type	Telco 1	Telco 2	Telco 3	Telco 4	Telco 5	Gateway GSM
Combined	VC1	USD 57,072.188	USD 67,880.919	USD 103,203.558	USD 41,250.712	USD 29,241.011	USD 19,842.115
	VC2 e VC3	USD 197,714.364	USD 224,848.083	USD 357,526.648	USD 251,439.128	USD 137,477.509	USD 75,974.413

If you execute the calls using the cheapest alternatives, you would use Telco 5 or use GSM gateways to transform calls that are fix-to-mobile into mobile-to-mobile.

In this way, you can calculate the minimum possible cost if you route the calls through the cheapest alternative. In this particular example, just doing that, you could achieve 27.95 percent savings over the current expenditures and do so without any tariff negotiation or implementation of GSM gateways.

Type	Least cost routing	Current cost	Savings	% savings
Fix-to-fix	USD 601,511.16	USD 663,442.00	USD 61,930.84	9.33%
Fix-to-mobile	USD 166,718.52	USD 402,829.00	USD 236,110.48	58.61%
Total	USD 768,229.68	USD 1,066,271.00	USD 298,041.32	27.95%

These kinds of calculations are simple, and you should be able to do them easily and regularly.

Very often, the difficulties associated with properly routing all traffic are linked to unavailability of trunks or routing devices. Nevertheless, you should be always aware of your potential gains if you had the resources.

Chapter 3: Guidelines for Planning a Call-Center Infrastructure

3.1 General rules

The first thing when defining the architecture of a telecom infrastructure of a call center is to identify the basic functions that the infrastructure is going to support:

- Switching
- Support to extensions
- Least-cost routing
- Dialing
- Call recording
- Number classification
- Automatic call distribution (ACD)
- Computer and telephony integration (CTI)

The designer of a call-center telecom structure can consider two extreme scenarios: only one device "box" executing all functions, and each function executed by a specific device.

The most common scenario is a structure between these two extremes, with two or three types of devices supporting the eight functions. Therefore, it is up to the designer to define the level of dispersion of the functions and which device is going to execute what (always remembering that you may have more than one type of device able to do the same functions).

As an example of a typical structure, we may have a PBX executing the switching, supporting the extensions, supporting the ACD and CTI functions, and executing the least-cost routing and a dialer executing the dialing, the call classification, and the call recording.

Concentrating functions in a few devices tends to make the structure cheaper and simplify the operation; however, very often it implies some sacrifice in performance.

In other words, devices that execute multiple functions very rarely do so with the same level of performance as devices with dedicated functions. Specialization tends to generate better performance. A general rule to guide your decisions about the distribution of functions in call-center structures is the larger the volume of the traffic and the more crucial the operation, the more extensive the specialization of the resources used (hardware and software) and the smaller the number of functions handled by each resource.

Of course, it is a general rule and should be judicially applied in a case-by-case basis. It is also worth mentioning that at one extreme of our possibilities we have dedicated hardware with only one function and at the other extreme general-function hardware (usually PC-based) supporting multiple functions.

Another important aspect to which the designer should pay attention is the issue of multiple technical environments and the need for standardizing these environments. Here the general rule guiding your decisions should be: the ideal is to minimize the number of environments and standardize these environments not only in terms of brands but, above all, in terms of functions per type of resource. Although a bit obvious, it is always good to remember that standardization tends to reduce training, maintenance, management, and support costs, and increases productivity. A good example of this kind of issue is demonstrated below:

Example 1:

In this example, we have a site with three distinct technical environments, each one of them deploying different hardware (from different providers) with the functions distributed differently by type of resource—a complete lack of standardization regarding not only brands but also regarding the function by type of device.

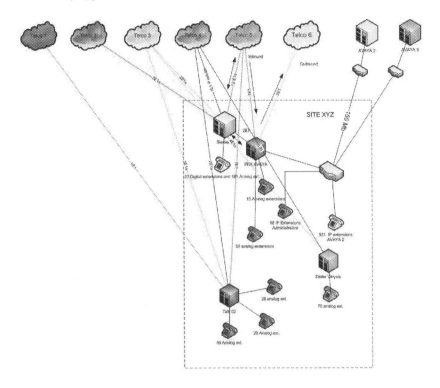

Environment 1 (Talk)—The same resource (dialer) dials, records the calls, routes the calls, and supports the extensions.

Environment 2 (Siemens and AVAYA)—The Siemens PBXs do the call routing; the AVAYA PBX supports the extensions and executes the dialing process; and a dedicated record machine does the call recording.

Environment 3 (Verysis)—The same function distribution as the Environment 1 but with different providers.

This is a completely nonstandard environment both in terms of brands and functions per type of resource. Clearly, it is not an ideal situation.

Example 2:

An intermediary example, much more homogeneous than the previous one:

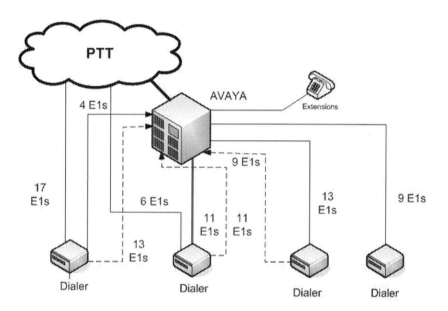

In this example we have a situation where the AVAYA PBX supports the functions of extension support, least-cost routing, switching, ACD, and CTI. The dialers execute the dialing, call classification, and call recording. Note in this example that some of the outgoing trunks of the dialers are connected directly to the public network. That indicates two possible scenarios: 1) part of the least-call routing is executed by the dialers, or 2) part of the traffic is not routed.

In any event, this example shows a structure much more stable than the previous one, although the concentration of functions points to the existence of problems for high-traffic volumes. In this example, we have a dedicated hardware (PBX) and generic hardware based on PCs (dialers).

Example 3:

Here we have an example of a structure more suitable for high traffic and critical environments.

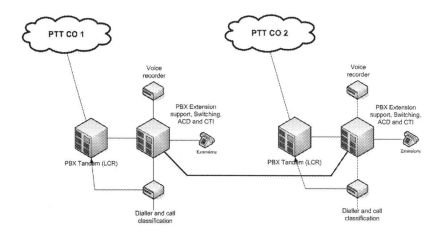

Note that this structure has two environments completely standardized, both in terms of brands and functions by type of device. Note also that the functions are broken down throughout more devices than the previous example. We have four types of equipment for eight functions. In the previous example, we had two for eight.

The function of least-cost routing is separated in one dedicated PBX and the extension support, switching, ACD, and CTI in the other. We also have the call recording done by a dedicated device.

Here it is interesting to notice that the separation in many technical environments (in the same site or not) should be done in such a way as to guarantee some level of interchangeability between them. It should be done in order to guarantee that, in case of problems with one environment, there is the possibility of handling at least part of the traffic through the other. (The deployment of IP extensions may make this process even more transparent). This would be what we could call "hot structural backup."

Typically, there is a correlation between the concentration of functions (using dedicated or nondedicated hardware) and multiplicity of technical environments. In general, we note an inverse correlation between the function distribution per device and the number of environments. This usually happens due to performance problems in environments with a high concentration of functions per type of device, which in its turn leads to the need for expansion through adding new environments.

In other words, organizations who opt for concentrated functions in a small number of types of devices tend to have performance problems early, which happens because the concentration of functions tends to limit the possibility of expansion within the same environment and consequently forces the organizations to expand through implementing new environments (very often not compatible with the existing one).

Without proper definition of how the functions will be distributed by type of device and when and how new technical environments will be implemented, such decisions end up being made by default, considering only the circumstances of the moment. Such a decision-making process may seem practical but tends to generate problems of performance, reliability, scalability, costs, and operational complexity in the medium and long terms.

Here it is worth emphasizing that planning does not imply buying super-dimensioned hardware—on the contrary. The existence of a plan allows everybody involved to know what to do when volumes reach some preidentified milestones. Scalability in this context is the key word.

When functions should be separated in dedicated equipment? When should new environments be implemented? Those questions should be preanswered in such a way as to guarantee a clear view about how the infrastructure will evolve when the volumes increases. The plan in this context is no more than a guide about what to do when the volume grows (with clear milestones).

A good planning process guarantees that the right questions be asked and answered in advance, avoiding unnecessary efforts spent on doomed initiatives. A good plan also helps when dealing with service and hardware providers, because it allows the organization to identify what it needs instead of merely reacting to what is offered.

When growth occurs through merger or acquisition, the telecom manager has to make good with what he or she gets, and in these situations you rarely have a lot of room for planning (at least until the deal is closed). But even in these scenarios, having a clear guideline about how the structure should operate is very important to help you standardize the environments (not regarding brands but surely regarding functions) and simplify the infrastructure. In these situations, it is always good to keep in mind the general rule that, usually, it is better to have good standardized environments than excellent heterogeneous ones.

3.2   Using IP extensions

An important aspect to be considered when planning a call-center technical infrastructure is the feasibility of deploying IP extensions, either through soft phones or through IP phones. The deployment of IP extensions opens the possibility of separating the sites where the calls are generated/received from the sites where the attendants actually operate. Although the marketing of this technology emphasizes the possibility of working from home, this dissociation opens promising possibilities for call-center operations, allowing it to concentrate its call generator/receiver centers in selected locations and to expand its attendance sites freely in other locations. Therefore, using IP extensions makes it possible to build a dialer center in sites well served by the public infrastructure and distribute the calls through a data network to several attendance sites.

This possibility is particularly interesting considering the fact that good locations for people (a central location with good transportation infrastructure) may not be equally good in terms of public telecom infrastructure.

In addition, the possibility of separating call generation and actual attendance opens the possibility of building infrastructures that can operate transparently as backups for each other. In this way, an organization could have two dialing centers and these sites could support each other transparently in case of problems in one of them.

Another important aspect associated with this technology is the possibility of subcontracting attendance sites without the need for a heavy investment in telco infrastructure. It allows the organization to outsource the attendance space and the only need in terms of telco infrastructure will be the data network.

3.3   Capacity planning and expansion

You must have a very clear process defining the planning of the infrastructure, and the infrastructure area must be an inseparable part of any discussions about new clients, new services, or expansion of the current ones. In addition, the infrastructure area must respond to any change in the demand or in the traffic profile (identifiable through monitoring or due to change in the processes).

The telecom area should anticipate those changes. By using the monitoring tools, the telecom area should be able to analyze if the resources available are adequate, and over- and under-capacity situations need to be identified before they adversely affect user services.

Telecom managers should also participate in the discussions about new services and new clients, indicating to the organization the impact in terms of telecom resources (technical and economic) of including or removing services. It is also crucial to have processes in place that give the telecom area some sort of insight about site expansions and organic growth. It should be done not in a sporadic way, but within a defined process. This process needs to involve the top level of management on at least an annual basis and must encompass the overall planning process.

Physical expansion of the telecom structure is often an interesting situation, as it is common for organizations to expand by acquisition. These are generally not announced even internally to an organization until a deal is done. In these cases, the telecom area needs to be part of the due-diligence function because a large acquisition could have a significant impact on both usage and cost base of the telecommunications infrastructure.

In practical terms, to be able to do the necessary calculations, you have to use the volume of traffic currently handled during the peak hour, during the peak day of the month, and, if it were the case, do it during the peak month of the year. You have to identify this traffic in terms of number of calls and duration in minutes, separating inbound from outbound.

Once you have this information, you must project the growth rate expected for a given period—let's say, the next twelve months. The time span defined depends on your organization's strategic plans, and somehow you must be made aware of them. Whatever the time span defined, you have to define the monthly growth expected and add it to the volume currently being handled. With this resulting volume, you will be able to do the necessary calculations.

Basically, you should follow two different methods to do the calculations: one for inbound traffic and another for outbound-Erlang C (based on retention) and Erlang B (based on denial), respectively. The different approach for the two different types of traffic are necessary because the treatment for each one is a bit different. For the inbound traffic, we expect to receive all the calls, even if we are not able to attend them at once, therefore queuing them for a while. On the other hand, when executing outbound calls, we assume that there is a limited space (number of outbound trunks) and if a number of calls above the number of available trunks are placed they will receive a busy tone (denial).

Note that the calculations used to measure the predictive dialers are Erlang C (retention). In practical terms, in a predictive dialer everything happens as if the dialer generates the calls (varying the dial rate based on the queue theory) and forwards them to the attendants. Therefore, the calculation is done as if we had an income inflow.

It is highly advisable to have separate trunk groups for inbound and for outbound services. This arrangement is important to avoid situations where calls can't come in because there are too many calls going out (or vice versa).

Here it is important to notice that, even when you have trunk groups with hybrid usage (inbound and outbound) in this scenario, you still have to calculate the in and out traffic during the peak hours, using both criteria and then add both values.

3.4   Evaluating new technologies

This is not exactly a process, but more like a group of activities, which nevertheless should be considered. It is a matter of due diligence of the telecom area to scan the market looking for products and solutions that may add value to the business of the organization. It is advisable that the technical people have access to magazines, conventions, or user groups where they can compare their own experience with other professionals. It is very important that the organization provides internal resources for this function, otherwise this will increase the influence of service providers and hardware vendors to the point that it is not in the organization's best interest.

Companies may have a structure that includes functional resources like telecom architects. These individuals should know what technologies are available and their applicability to the organization.

The telecom area is also responsible for training the users of the telecom resources, letting them know how to make the most of the available resources.

Call-center operations usually are very dynamic environments, and expansions, moves, and shrinkages are extremely common. Therefore, being able to measure properly the devices used is part of the day-by-day work of any call-center manager.

3.5   Maintenance

At this point, it is important to mention that the organization has to have a policy regarding maintenance. In general, in large installations it is highly advisable to have permanent maintenance contracts. The absence of maintenance contracts exposes the organization to downtimes, even if there is the possibility of paying the hardware provider on a per-visit basis. It may result in a lack of a high-priority response in a moment of need. Usually, large and critical operations not only have permanent maintenance contracts but these contracts include very strong SLAs and preventive checkups. As mentioned, the need for permanent maintenance contracts grows as the size, criticality, and complexity of the infrastructure grows.

When contracting maintenance, some issues should be observed:

1) The basic reference in defining the monthly price to be charged is the value of the equipment. In this context, the price paid monthly corresponds to a kind of insurance. The normal price usually spins between 0.5 percent and 1.5 percent.

2) The maintenance provider must have all necessary hardware for ready substitution in case of need (passive reserve). Eventually, a cheaper contract may be negotiated transferring the responsibility of keeping backup hardware to the organization, which is cheaper but not advisable and only very large organizations may use this tactic.

3) Any contract must include a periodic configuration backup. The backup should be stored outside the organization by the service provider.

4) Precise SLAs should be defined, indicating the maximum response time, including diagnosis and fixing of the problems. There should be a maximum resolution time after a maintenance request. The contract should also indicate exactly how this time is to be counted.

3.6 Using Points of Presence (POPs)

3.6.1 Inbound operations

Many inbound call-center operators outsource their transport services, relying entirely on a public-dialed network to provide the interconnection between their clients and their attendance sites. This strategy, although practical in operational terms, very often results in higher transport costs. The reason is that outsourcing the call transport entirely deprives the call-center operators of some fundamental possibilities for reducing operational costs:

- Usually, the 1-800 contracts don't discriminate the origin of the calls, defining a flat rate per spoken minute (very often with discount per quantity) regardless from where the call was placed. Not having a distributed traffic capitation structure (POPs) makes impossible for the call-center operator to take advantage of the tariff system by exploring different interconnection costs associated with the users' geographical distribution. For example: in Canada, the calls within the same area codes are not charged; that means if you only have a 1-800 number and are locked in to a flat-rate contract, you will pay for the call even if most of your clients are calling from the same area code as your attendance site.
- Secondly, if a call-center operator doesn't have any capitation node (POP), it cannot distribute its IVRs and therefore cannot take advantage (in terms of transport costs) of the services whose process can be addressed by these devices, which obligates all calls, including the ones to the IVRs, to be transported all the way to the attendance sites.
- In some countries there are services (like 4000 and 0300 in Brazil) where the cost of the call can be divided between the callers and the call-center operator. In these circumstances the client usually pays for the local access and the call-center operator pays for the long distance part of the call—that means from the client's perspective the access is just a local call. By having a POP in these locations the call-center operator can take advantage of this cost structure (avoiding the long-distance cost).

- In countries where local calls are free or relatively cheap when compared with long distance, it may make sense, depending on the volume, to have POPs to collect the traffic and bring it to the attendance sites.

Therefore, if the call-center operator decides to implement its own transport network, usually there is plenty of room for optimizations. But not just any design will bring savings. An overengineered and badly designed transport structure may become even more expensive than an outsourced one. Consequently, the challenge is to design a structure that minimizes the costs involved with transporting the calls while maintaining the same level of service.

The magnitude of the savings achievable by implementing this strategy varies and is directly related to the geographical dispersion of the users. However, through the careful analysis of traffic flows, interconnection costs, and tariff rules, it isn't unusual to find over 50 percent savings in real dollar terms over the completely outsourced transport structure.

3.6.2   Outbound operations

The same logic as applied to inbound calls can be applied to outbound calls. Having a network of POPs allows you to transport the call to the nearest POP using the public infrastructure only between the POP and the number called. Depending on the tariff system, this strategy may save you a huge amount of money. For instance, in some countries, as already mentioned, there is no charge for local calls. That means, depending on your long-distance cost and the percentage of calls within the metropolitan areas, you may save most of your telco cost if you have a network of POPs. Although it is particularly true where local calls are free, this idea also can be applied anywhere else. The basic logic behind this is that the cost of the network of POPs, plus the cost of the local calls, may be lower than the cost of the long-distance calls.

Therefore, analyze traffic and then map where it would be economically feasible to have a POP. This can be a very interesting option, depending on the case. In addition to actually building the structure of POPs, this kind of analysis can be used as reference data when negotiating long-distance tariffs with telcos. You can say, "Give me this price, otherwise I will transport my calls through my own network."

Subsequently, we are going to discuss a case in which the alternative of having a private transport network was the deciding factor in the negotiation process.

3.7 Mapping the infrastructure

Here we are going to describe some additional information, and its several aspects, that you should strive to have. This information is instrumental for telecom management in a call center and as such should be carefully controlled.

The first set of basic data is the site locations and the services operated in each one of them. This compilation has to include also the number of minutes in and out per month (including historical data) and the percentage of the total they represent. Although it may seem a bit obvious, this can be very complicated to keep control of in large multisite operations. This control should also include the Erlangs demanded and the number of trunks available per service provider. The spreadsheets below illustrate the kind of view the telecom manager has to have:

Site	Minutes month	%
Site 1	4,848,631.82	37.45%
Site 2	8,097,991.29	62.55%
Total	12,946,623.11	

Or a bit more detailed:

Site	Minutes	%	Day of maximum traffic (FCT- day = 6,85%)	Hour of maximum traffic (FCT - Hour = 10,86%)	Erlangs HMM
Site 1	4,848,631.82	37.45%	332,131.28	36,069.46	601
Site 2	8,097,991.29	62.55%	554,712.40	60,241.77	1,004
Total	12,946,623.11		886,843.68	96,311.22	1,605

Based on the required Erlang and using Erlang calculators (see chapter 11 on Erlang calculators), we calculate the number of necessary trunks (and E1s or T1s needed):

Site	Trunks	E1
Site 1	630	21
Site 2	1034	35
Total	1664	56

Here we can easily calculate the adequacy of the number of trunks really available and those needed, considering the traffic actually supported.

The second set of information is linked with the details of the equipment located in each site. This identification should be done through what we call an "availability sheet," which is a spreadsheet indicating the details of the equipment, including the resources in use and the resources available.

Equip name	In use					Equipped					Available				
	Analog extensions	Digital extensions	IP extensions	Analog trunks	Digital trunks	Analog extensions	Digital extensions	IP extensions	Analog trunks	Digital trunks	Analog extensions	Digital extensions	IP extensions	Analog trunks	Digital trunks
Client 1	84			300	210					360					
Client 2	76														
Client 3	2														
Client 4	3														
Client 5	5														
Client 6	3														
Totals	173	0	0	0	300	210	0	0	0	360	37	0	0	0	60

This spreadsheet should be updated each time any change is made in the infrastructure and should be used as a resource inventory control. This control should include equipment such PBXs, dialers, recorders, IVRs, etc. It is through this spreadsheet that you verify the availability of resources to subsidize the sales people or internal users, indicating what you have readily available and what would take time to provide.

This kind of control should also include the trunks, indicating the number of trunks of each service provider in each address. The spreadsheet below exemplifies this kind of control:

Provider	Site 1 Total (E1s)	Site 2 Total (E1s)	Total (E1s)
Provider 1	58	0	58
Provider 2	19	70	89
Provider 3	0	28	28
Provider 4	0	1	1
Provider 5	0	8	8
Provider 6	0	30	30
Total	77	137	214

The third set of basic information is the diagrams of the sites indicating the equipment and how it is connected. Each site must have a detailed diagram, such as the following:

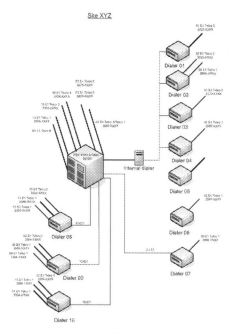

Note in this particular example that the trunks linked with the public network are connected directly to the dialers and the PBXs function only to support the extensions and do the switching. This particular interconnection strategy may have some not-so-obvious drawbacks:

1) It makes difficult least-cost routing, because any change of routes has to be configured in each dialer individually.

2) It doesn't allow the pooling of groups of trunks per service provider that forces each dialer having trunks of each service provider. The consequence is a situation in which the number of trunks in each dialer for each telco may be smaller than is necessary at some moments, which usually provokes overflow through trunks of service providers with higher tariffs. The alternative would be to install larger numbers of trunks in each dialer, which would be empty most of the time and would increase the overall hardware cost.

3) It makes the billing process more complex, because it forces you to collect the logs from each dialer and be alert for duplications.

Chapter 4: Methodology to Map and Analyze the Traffic in a Call Center

In this chapter, we are going to describe a methodology that allows you to establish the optimal correlation between the users' geographical dispersion, the traffic volumes/flows, and the tariff system. This kind of analysis provides you with a complete view of the trade-offs between cost, performance, and reliability, and tends to produce much more effective and lasting savings.

Therefore, getting the right design is important. Overengineering an overpriced structure does not require methodologies, tools, or elaborate processes. However, the true challenge lies in designing an optimal structure, which minimizes cost while maximizing performance. To achieve this objective without an analytical method and executing all necessary calculations manually is virtually impossible, and that's why a methodology and algorithms are so important.

The magnitude of the savings achievable using this strategy (mapping traffic interest, volumes, and profile) varies and is directly related to the size and geographical dispersion of the users and the organization's sites; nevertheless, the savings usually are substantive.

The same process described here, in addition to carefully crafting the structure design, also assists in several other aspects associated with contracting, pricing, and managing telecom in a large call center. These include activities such as:

- Evaluate service providers' bids, comparing fairly, different kinds of services, technologies, and pricing strategies.
- Evaluate how much would be fair to pay to outsource a transport structure.
- Analyze the current prices, comparing them to the several interconnection alternatives available throughout the market (benchmarking).
- Evaluate the feasibility of integrating voice and data using different strategies and technologies (for example, integration—total or partial—using VoIP, Voframe, or VoATM [Voice over IP, Frame Relay, or ATM]).
- Negotiate telecommunications budgets by establishing a clear correlation between traffic, quality of services, and costs.
- Simulate future needs and verify how the structure cost will behave faced with increase in traffic (assisting in strategic planning and anticipating needs and problems).
- Analyze traffic and check if the number of trunks, the bandwidth allocation, and CIR (Committed Information Rate) definitions are adequate (capacity planning).

At this point, it becomes clear which cost factors in the actual structure could be reduced, and it becomes possible to produce a very straightforward, high-level management report comparing the actual expenditures with the proposed ones. The necessary investments and potential savings can be clearly identified, and the return over investment of the project calculated. Those analyses are crucial when looking for funding.

In addition, it becomes possible to properly analyze alternatives, such as outsourcing or external management. Knowing the cost to build an optimized structure better positions an evaluation of cost benefits of various outsourced solutions.

Although the identification of an ideal structure to support a given traffic volume is in itself a huge benefit, the ability to calculate these structures quickly allows the organization to perform many calculations using several traffic volumes and to establish the correlation between volume and cost. Such calculations allow the organization to project the growth of the traffic, verifying not only how much the structure could cost today but also how much it would cost if the traffic increased by a given rate.

Here we are going to discuss a methodology to map and optimize traffic in call-center structures. This is a methodology applied to the specific needs of large call centers that support populations over wide areas and with multiple service offerings.

The objective is to design a structure that minimizes the costs involved with transporting the calls through the careful analysis of the traffic flows, interconnection costs, and the local tariff rules.

The whole process pivots on trying to make the call-center structure as compatible as possible with the users' geographical dispersion and interconnection costs (tariff system).

Through the methodology and tools described here, you can identify the correlation between traffic volumes, infrastructure cost, and revenues as they relate to the services offered by the call centers. For example, assume that a business is considering a new service offering one or all of the call centers in a country. The analysis described enables the modeling of business cases that identify the correlation between each new call-center service and the associated cost involved in implementing it effectively.

We are going to describe the process phase by phase, and it does not require the reader to be an expert in telecommunications. However, some knowledge of the technical terms and functions/features available in the different devices is beneficial for the overall understanding.

4.1   Overview

The methodology to design call-center structures follows three phases:

- Data gathering/files preparation
- Ideal structure identification
- Refining the results

These three phases are described in more detail in the following subsections.

4.2   Data gathering/files preparation—Phase 1

The first phase consists of identifying and formatting the information necessary to perform the calculations associated with identifying the best designs. The information is grouped into five categories:

- Services provided by the call center
- Points of origin/destination of traffic
- Interconnection costs
- Hardware specifications and costs
- General network parameters

The objective is to identify the locations, volumes, and flows (the traffic matrix—see chapter 11) and calculate the costs involved in the transport of these flows.

Given this information, analytical tool software can run all possible combinations and select the most cost-effective solution, depending on the minimum requirements in terms of performance and resilience (i.e., redundancy). It is also very important to note that our results will only be as accurate as the data used to feed the algorithms.

Experience shows that, in some cases, generalizations in grouping the users can be made without damaging the results; another generalization can be made using the same interconnection costs when analyzing structures within the same area.

Services provided by the call center

The first step of the first phase identifies all services provided by the call center. Here we identify the services: volumes (in/out) and corresponding live-attendant locations.

The calculations can identify the ideal number of live-attendant sites and their ideal locations. However, if for some reason we don't want this identification done automatically, we can predefine it while taking into account the existing infrastructure. That means we may be able to calculate the best locations for our attendance sites, or we define their location arbitrarily (if the relocation of sites is out of question).

The volumes here should be represented in terms of the number of calls per user per month, the duration of these calls in minutes, and the hourly distribution. In addition, we also have to identify parameters such as the percentage of all potential users that really use the service, the percentage of the calls that demand live attendance, and the percentage that can be completely covered by the IVRs (Interactive Voice Response).[1]

Here it is important to explain that in most cases you don't have such precise information about each service and each user. Usually, it's possible to identify the volumes per area code and, based on that, do the calculations. What was explained in the previous paragraph was conceptual. In other words, in theory, we depart from the client's locations.

Points of origin/destination of traffic

The second step of the first phase is the identification of all locations where traffic is received or originated. Points of origin and destinations of traffic show the users locations and the possible (or predefined) attendance and caption-node locations.[2]

When designing a call center, people are our source/destination of traffic. Therefore, theoretically, to calculate these structures we would need to identify the location of each individual. However, in the real world this approach isn't feasible. Actually, what we normally end up doing is grouping the sources/destinations population by some criteria, such as

---

[1]   For a more detailed description of how to identify the volumes, please refer to the spreadsheets shown in chapter 2 on traffic management.

[2]   Caption nodes are nodes that aggregate local calls and provide services to the client. They may also be described as backbone nodes or POPs.

area code, zip code, CO (Central Office—Telco agregation hubs to where the local loops are connected), and process it grouped. You should make assumptions that allow you to define more precise origin/destination criteria.[3]

Examples:

- You cannot pinpoint the origin of each individual call to a bank call center, but you can assume that most users will originate their calls from the same area code as their branches, at least most of the time.
- When analyzing a call-center structure for a computer manufacturing company, you don't know with 100 percent certainty from where each request for technical assistance will originate, but based on the sales distribution per geographical area, you can create a reasonable model.
- When designing a call center for a credit-card company, you can group its clients based on their home location area code. Of course, not 100 percent of them will originate their calls from there, but it is reasonable to assume that the majority of them will.
- When designing an outbound call center for general sales, you may use the general geographical distribution of the population (separating a percentage of it as real users) to base your calculation on.

Information such as the population living in the areas and the known correlation between population size and calls are basic parameters when analyzing a call center (population here means the potential users, who eventually can be the whole population or particular segments of it, such as the clients of a specific bank). In general, having this information you can infer a fairly accurate model.

---

[3] Of course, this kind of verification has to be done when you plan a new call center. Usually, when dealing with an existing center, the job becomes simpler because you can just group the calls by area code as described in the sub-item "traffic interest."

Regardless of all the difficulties involved when defining points of traffic origin or destination, usually you can manage to establish a reasonable model correlating the traffic sources/destinations and its volumes.

Interconnection costs

The third step of the first phase identifies all the interconnection costs. When analyzing a call-center structure, having a complete view of the interconnection costs is absolutely critical. You have to know how much it is going to cost to transport your data/voice flows. If you are not reading this book piecemeal, here you should review "Benchmarking" and "Interconnections costs" discussed in chapter 1, "Cost Management."

Usually the connection between the users and the caption nodes (known as access nodes) happens through the telcos who charge a predefined and widely known rate per minute. However, between the caption nodes and the live-attendant nodes you have a backbone, which can usually be provided by a variety of service providers using many technological alternatives.

Usually, you integrate voice and data over this backbone. The voice flows are the digitalized and compressed calls between the users, and the live attendants and the data flows are the IVRs requests to/answers from the company's hosts (if the IVRs are distributed throughout the POPs).

Another important factor is the availability of interconnections among the potential nodes (caption and live-attendant). In an ideal world, all nodes could be connected to each other, but experience shows that in many cases, such as when there is an existing public infrastructure, the options are limited. Even if you are willing to build an entire private infrastructure, some routes are not realistically feasible.

The first step is to identify the available interconnections. The next step is to identify the rules that define the interconnection costs. Then, the costs are calculated using *interconnection cost calculators*. By following these steps you can identify the costs for any given connection for any given bandwidth, using all available service providers and technologies.

Luiz Augusto de Carvalho and Olavo Alves Jr.

4.2.1 Interconnection-cost calculators

Interconnection-cost calculators are tailor-made programs that can identify the cost to interconnect two sites by using a particular service provider and particular kind of technology for any given traffic volume.

Interconnection costs can be represented in several different ways as can be seen below:

Example:

- Carrier lease lines; typically represented as a function of distance × volumes.
- Fiber-optic infrastructure; usually cost is a function of distance.
- Dial-up access; normally follows a PSTN price spreadsheet widely known or predefined flat rates per minute (for example: 1-800 agreements).
- Packet networks; typically follow a function correlating CIR × Volumes.

In most cases, you don't need to have the costs of each possible interconnection to produce a credible analysis. You infer general rules, based on limited information. For instance, most service providers never provide their clients with the tools to calculate the prices of their services by themselves, preferring to produce proposals on a case-by-case basis. Faced with this kind of limitation, it is perfectly acceptable to infer the general rule from a relatively small number of samples using nonlinear regression techniques.

In large structures with many interconnection provider alternatives, you can create more than one interconnection cost calculator. As a result, you can identify the ideal solution by defining which technology and service provider tends to be more suitable for each interconnection or type of traffic (least-cost routes).

This information is crucial for calculating the most cost-effective call center. If you are a call-center designer, you may be able to reuse this kind of information for other projects within the same geographical region.

Hardware modularity and costs

The fourth step of the first phase identifies all hardware modularity and costs. In order to perform this identification, you have to understand that in a call-center design process the devices involved can be divided into two groups: caption-node equipment and attendance-site equipment.

Of course the equipment cost will vary significantly, depending on how many caption and attendance sites you use and how this equipment will be interconnected. The hardware costs should be identified based on technologies and volumes.

In summary, a clear method of identifying the load/cost curve of your hardware is essential. The application should be able to answer questions like the following: "How much will my hardware cost if I have ten thousand users connected to this node, and how much will it cost if I have one hundred thousand users instead?" We have to identify these rules and insert them into the software (*equipment cost calculators*).

4.2.2 Equipment-cost calculators

Equipment-cost calculators are tailor-made pieces of software that can calculate the cost of each type of hardware based on a given traffic volume. This volume can be number of people or total traffic volume in/out flowing through the node.

The first step when defining the hardware-cost calculator is to identify which equipment is going to be deployed by the nodes, defining what we call "a typical node." The second step is to establish the interrelation among these various pieces of hardware.

Example of hardware-cost calculator.

Here we will simulate a very simple cost calculator, just to give the reader a general understanding. Real equipment calculators usually are much more complex and involve many other details. In our example, we are designing a call-center structure to collect inbound traffic. Below you can see a typical inbound node with the IVRs (Interactive Voice Response) devices decentralized:

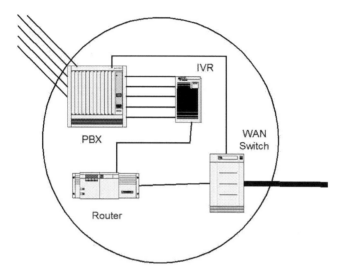

In this example, our typical node includes four types of equipment:

- PBX
- IVR
- Router
- WAN Switch

All equipment size is based on the number of calls received (this particular example is an inbound-only call center). With the number of calls, we calculate the number of trunks necessary in the PBX. The incoming calls are divided between the ones attended by the IVR and terminated there and the ones that demand live attendance, which can only be provided for the central node. Therefore, these calls are transferred through the PBX digital trunks and through the WAN switch to the central node.

The IVRs also demand contact with the central node in order to access the company's data centers, and therefore part of the bandwidth will be used with this data traffic. Based on this process we can write a code as follows:

```
* PROGRAM : PRICEHAR1.prg

* WRITTEN BY: Luiz Augusto de Carvalho

* DATE : 06/02/2002

* OBJECTIVE : Call-center hardware calculator

************************************************************************

* Calculating PBX cost

************************************************************************

PBXhack=10000 && PBX rack supporting 20 cards into two shelves

Extensioncard=1000 && With 32 extensions each

analogcard=2400 && With 16 Analog trunks per card

digcard=1800 && Digital trunks 30 trunks per card 1 E1


ntrunks= 200

ncards=0

* PBX—Calculation *

* Analog trunks *

************************************************************************

ntrunkcard=int(ntrunks/16)+1


Ncards=ncards+ntrunkcard


ptrunkcard=ntrunkcard*analogcard

* Digital trunks

************************************************************************

ndigtrunks=int(ntrunks*0.6)+1 && 60 percent calls need human attendance


if (ndigtrunks/30)-int(ndigtrunks/30)<>0

ndigcard= int(ndigtrunks/30)+1

else
```

```
ndigcard= int(ndigtrunks/30)

endif pdigcard=ndigcard*digcard

Ncards=ncards+ndigcard
```

* Extensions *
****
```
nextension=Ntrunks

if (nextension/32)-int(nextension/32)<>0

nextcard= int(nextension/32)+1

else

nextcard= int(nextension/32)

endif

pext=nextcard*extensioncard

ncards=ncards+nextcard
```

* Magazine PBX *
****
```
if (ncards/40)-int(ncards/40)<>0

nhack= int(ncards/40)+1

else

nhack= int(ncards/40)

endif

phack=nhack*pbxhack pbxprice=phack+ptrunkcard+pdigcard+pext
```

* IVR Calculation *
****
```
pivrhack=800 && IVR hack cost with 6 slots for cards each

ivrtrunk = 70 && Card with 8 trunks each
```

```
nivrcard=int(ntrunks/8)+1

if (nivrcard/6)-int(nivrcard/6)<>0

nivrhack= int(nivrcard/6)+1

else

nivrhack= int(nivrcard/6)

endif

pivr=(ivrtrunk*nivrcard)+(pivrhack*nivrhack)


? pivr


* WAN switch Calculation *
***********************************************************************************
pcard1=1500

pcard2=500

switchhack=20000

nswitch=ndigcard+1

nswitch=2*nswitch

if (nswitch/16)-int(nswitch/16)<>0

nswitchhack= int(nswitch/16)+1

else

nswitchhack= int(nswitch/16)

endif

pswitch=(switchhack*nswitchhack)+(nswitch*(pcard1+pcard2))

? pswitch


* ROUTER Calculation *
***********************************************************************************
router = 2500 && router CISCO 2509 US$ 2500
```

? "Total price = ", str((pbxprice+pivr+pswitch+router),10,2)

* montly rate .05 including maintainance

? "Total monthly price = ", str(((pbxprice+pivr+pswitch+router)*.05),10,2)

With careful analysis, we build detailed hardware-cost calculators, and depending on the complexity of the project, we can even include many possible vendors. Including more than one possible vendor allows us to determine both the ideal structure and the preferred vendor.

The results calculated by the hardware-cost calculators should provide monthly based costs, covering the complete ownership cost, including equipment, maintenance, and management. We have to decide which percentage of the cost will be considered. Typical values are:

Monthly expense	%
Monthly maintenance	1%
Monthly Management	0.5%
Monthly 5 years leasing	3%

These financial parameters have to be defined when preparing the hardware-cost calculator and should be identified in the market where the equipment will be ordered/deployed, being as close as possible to the real market practice and the WACC (Weight Average Capital Cost) of the organization.

4.2.3 Nonexclusive information (including general parameters)

Nonexclusive information includes:

- All information that can be used as valid reference in the absence of specific data
- Parameters defined, based on our understanding of what is the ideal quality of service

Therefore, depending on which specific information is available to execute a project, we may need to use more or less nonexclusive information. For example, if we don't have precise measurements or it is a completely new installation, we can use traffic patterns of organizations with similar operational patterns.

By the same token, if we don't have specific quotations for interconnections or hardware we may use nonexclusive information gotten through quotations done in other organizations within the same area. This type of possibility is what drives the sales of services of specialized consulting companies. They are able to provide typical parameters and nonexclusive information to bridge information gaps.

The other category of nonexclusive information is the parameters. We have to identify all general parameters and strategies necessary to calculate the call center properly. Such parameters include a wide range of characterisitcs from failover strategies to buying policies. We divide these parameters into three basic groups:

- Strategic
- Technical
- Operational

Some of these parameters can be defined on an individual basis, but usually we adopt standardized definitions for the whole project. For instance, we can define that our acceptable loss is 1 percent for all kinds of services instead of adopting individual values for each one.

Examples of strategic parameters:

- Fail strategies
- Equipment functions and interconnections strategies
- Acquisition policy (buy, leasing, or rent)
- Type of 1-800 agreement/access paid by the client or by the company
- Insource or outsource of telecommunications services

Examples of technical parameters:

- CIR range
- IVR placement
- Access-traffic splitting
- Voice derivations to/from IVR (percent)
- Quality of Services parameters (latency, transference rate, voice compression rate, loss)
- Traffic priority
- Path length (number of acceptable hops)

Usually, we execute the calculations using a voice-loss pattern of 1 percent and overhead for IP encapsulating of 5 percent.

Examples of operational parameters:

- Maintenance policy
- Network management policy

When identifying these parameters, it is common that unrealistic expectations about the quality of service desired are set; most times, these unrealistic expectations are motivated by the lack of understanding of the correlation between cost and performance.

The definitions of the adequate quality of service of each application have to be based on the requirements of each application used by the organization.

We need to have a clear view about the implications in terms of costs and benefits when defining the parameters.

Confronted with these problems, we can adopt several strategies from precise measurements to reverse calculation based on nonexclusive information (inferences). The ideal strategy is specific to each situation.

4.3   Generating the designs—Phase 2

Once we have the necessary data, it is time to design the call-center structure. In this section, we will discuss how to execute an effective design. Here it is important to keep in mind that arriving at a design that works is important; however, it isn't enough. Designing an overengineered structure isn't difficult. The true challenge lies in the design of an optimal structure, which minimizes cost while maximizing performance.

As already discussed, the problems faced by a call-center designer are basically logistic problems. For instance, most organizations need to find the best possible location for their offices, manufacturing sites, and warehouses. A bad location will be translated into higher costs and lower competitiveness. The same concepts apply to the call-center structures—bad logistic arrangement will generate higher costs.

We will also describe in this section the concepts involved in the design of call centers. It can be done manually or deploying calculation tools. However, in our view, designing properly a large call center by doing all the necessary calculations manually isn't practical given the fact that it would take too much time and would increase the possibility of mistakes beyond a reasonable standard.

The algorithms deployed by these tools are public knowledge, and it isn't our objective to discuss the codes of such algorithms but rather the concepts behind them and how to deploy them (the concepts described in this book can be implemented in several ways). To get samples of such codes see the bibliographic references. We are going to describe briefly the several classes of algorithms and the problems addressed by them.

Design algorithms can be divided into several classes, depending on the type of problem. We can consider basically the following classes:

- MST (Minimum Spanning Tree)
- CMST (Constrained Minimum Spanning Tree)
- Smaller/cheaper route
- Node-location identifiers
- Site associations

- Optimum routing
- Capacity planners
- Meshed networks
- Interconnection-cost calculators

Once again, it is important to understand that this book deals with the logic supporting the algorithms not the algorithms themselves. Another important factor is that such analyses demand a thorough data gathering, and only after getting the necessary information can we initiate the analytical process. The analytical process typically encompasses two phases:

- Topological scenarios generation
- Testing all topological scenarios using the interconnection costs available

All the tools presented in the bibliographic reference section will follow the logic described here, and we recommend reading this chapter before trying to understand the codes themselves.

The process of identifying the ideal call-center structure is based on the following concept: a call-center capitation network can vary from a completely centralized structure (star topology) to a completely distributed one, as you can see in the example below, which gives a conceptual view about how the creation of the topological scenarios occurs.

Example:

Here we have a capitation structure with twenty remote area codes, each one with around one thousand users and one attendance site. When analyzing the structure, we vary the number of users, which makes the existence of telecommunications aggregation nodes feasible from one thousand to twenty thousand. Doing that, we identify all possible topologies, and together with the interconnection costs, we can calculate which topology produces the cheapest design.

1) One thousand users are enough to justify the creation of a telecommunications aggregation node.

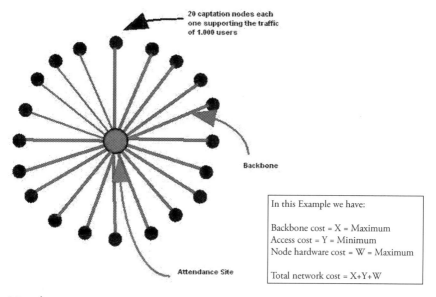

In this Example we have:

Backbone cost = X = Maximum
Access cost = Y = Minimum
Node hardware cost = W = Maximum

Total network cost = X+Y+W

20 nodes

2) Ten thousand users are enough to justify the creation of a capitation node.

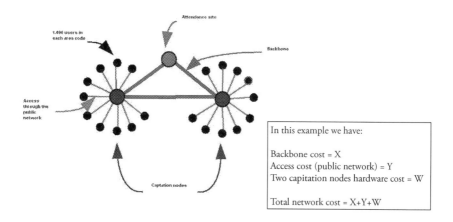

In this example we have:

Backbone cost = X
Access cost (public network) = Y
Two capitation nodes hardware cost = W

Total network cost = X+Y+W

2 Nodes

3) Twenty thousand users aren't enough to create a capitation node.

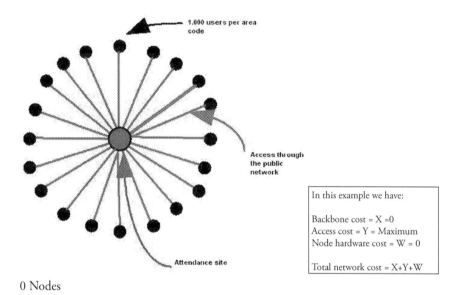

0 Nodes

As can be seen in the examples above, there are three basic elements in a call-center structure:

- Access using the public-dialed network, defined as the connection between the users and node (capitation or live attendant)
- Backbone, defined as the capitation node interconnections
- Capitation-node hardware

The most cost-effective structure will be the one in which the sum of these three factors is the smallest possible:

The logic of this analysis is shown in the formula:

ACCESS COSTS + NODE HARDWARE COSTS + BACKBONE COST = TOTAL COST

As you can see, the access cost tends to be inversely proportional to the number of capitation nodes. It means that, the more capitation nodes we have, the cheaper the access cost tends to be. In the other hand, the more users we have linked to a capitation node, the bigger the node's hardware cost and the backbone cost tend to be.

Here it is important to understand that this is what we call a "heuristic problem"—that means a problem to which there is more than only one possible solution, but rather there are several. This is why operational and management factors are so important to complement the process of choosing the best design.

The challenge is to identify which combination of these three factors will produce the most cost-effective solution. We find this ideal point through exhaustive search, calculating all possible scenarios, and going from the complete star topology where the backbone cost is zero to completely distributed structure in which the access cost is minimum.

In other words, we define that 1 percent of the users make a capitation node feasible. Then, we vary this value from 1 percent to 100 percent to identify all possible scenarios. Proceeding in this way, we identify the number of capitation nodes for each amount of users (scenario). Therefore, we have all scenarios between these two extreme situations:

One user is enough to make a capitation node feasible (dedicated connection to the attendance site)

- Dialed access cost minimum and backbone cost maximum

On the other extreme, if the number of users required to make our node feasible is equal to the total number of users (100 percent), our call center will be completely centralized:

- Dialed access cost maximum and backbone cost minimum

Usually the optimum scenario is somewhere in between these two extremes. The second phase is divided into two sub-phases: clustering and alternatives comparison.

4.3.1   Topological scenarios generation

The first step is to determine where and how many aggregation nodes there are (backbone definition). In this phase, our network will contain two distinct groups of structural entities: the aggregation nodes (backbone) and the traffic units (users/group of sources of traffic).

Using a hybrid algorithm such as Prim-Dijkstra, the backbone nodes will be connected among themselves. The *traffic units* (understanding traffic unit as a user or group of users) will be connected to the nearest node in a star topology (access).

As already mentioned, the topology-generator algorithms work by defining which traffic volume makes the existence of an aggregation node feasible, varying this volume from a minimum amount (a user, for example) to the total call-center traffic volume. Proceeding in this way, they manage to generate all topologies from a completely distributed to a completely centralized structure. If, for instance, we assume the minimum volume to justify an existence of a capitation node is the traffic generated by a traffic unit (assuming all traffic units have the same traffic volume), we can easily perceive that the topology will be completely distributed and the cost situation will be access cost minimum and backbone cost maximum.

On another hand, if the number of users/calls to make a node feasible is equal to the total number of traffic units (users/calls), the capitation network will be completely centralized with access cost maximum and backbone cost minimum.

At this point, the objective is just to identify the possible topologies. Therefore, we identify how many and where the capitation nodes are in each scenario (understanding a scenario to be the volume of traffic to make a capitation node feasible).

The aggregation-node identification is made based on a predefined group of possible nodes, distance, and traffic volume. The potential nodes can be arbitrarily defined—the distances are calculated based on the geodesic coordinates and the volumes based on the *traffic pattern* of a typical *traffic-generator unit* (users or group of users) and the number of *traffic-generator units* per *traffic unit* (users per area code for example).

It is important to understand that the identification of all possible topological scenarios is just the first step in the process of designing a capitation structure. The ideal structure will be identified only after testing each possible topology and considering all available interconnection costs.

4.3.2   The process of identifying the topologies

Center of Mass algorithms (COM) and Sharma algorithms are examples of methods which can be used in this kind of analysis.

The algorithms to identify the aggregation nodes have as their objective to, given a group of possible nodes, identify the number and location of the nodes based on the minimum amount of traffic that would make a node justifiable.

Basically, the process consists of calculating the distance between each *traffic unit* and all possible nodes. Once this is done, you can verify the smaller distance and associate the volume of the traffic unit to the potential node. We repeat this procedure for all *traffic units*. When we finish this process, we verify the volume of traffic associated with each potential node. The potential nodes whose traffic volume equals or exceeds a given value are saved in this topology.

Then, we vary the minimum volume of traffic (to justify the existence of an aggregation node) and repeat the procedure. Usually we initiate the calculations defining the initial minimum volume and defining the incremental volume. Typically, we define as the initial minimum volume the volume generated by the *traffic-generator unit* (typically a group of users within a defined area code).

This process generates a range of topological scenarios from a completely distributed to a completely centralized topology.

In the end of this process, we have a file with all nodes identified considering all possible volumes (volumes to justify the existence of a node). The sequence of activities executed by these algorithms vary but usually they are:

1) Calculate the traffic load (in/out) of each *traffic unit* (usually users or calls per area code) based on the number of *traffic units* and in the *traffic pattern*.

2) Calculate the smaller distance between each *traffic unit* (area code, for example) and the possible aggregation nodes (predefined) identifying the nearest aggregation node.

3) When finishing the distance calculations, verify how many *traffic units* were associated to each node and the total associated traffic.

4) Once the traffic volume associated with each node is verified, then verify if this volume equals or exceeds the volume of traffic defined as the minimum volume to justify a node. If the associated volume is smaller than the minimum, the node is eliminated as a potential aggregation node. The process is repeated until all nodes have a volume equal or superior to the defined minimum value.

5) When all nodes have a volume equal or superior to the minimum volume defined, save the topology. Then,increment the minimum volume and repeat the steps 2, 3, 4, and 5.

6) Repeat this process until the minimum volume to justify a node is equal to the total call-center traffic. Then you have all possible topologies.

4.3.3   The traffic volume calculation of each traffic unit

In this process, you do the calculations based on the number of *traffic-generator units* (in a call center that means users—in or outbound) located in each *traffic unit* (typically area code).

The calculation of this load is crucial, because the clustering process is based on distance and load. The aggregation-node load in its turn is identified, based on the sum of the loads of the *traffic units* associated with the node. Based on these calculations, you identify the traffic in/out of each aggregation node and build the traffic matrix. For example:

		21	11	512	246	123	852	482	922	192	244	612	473	912	152
21	RIO DE JANEIRO	26879	95604	48597	1941	804	3538	1976	3448	3063	652	7780	993	2603	392
11	SÃO PAULO	21148	578	2686	84	611	580	402	597	1267	5	1629	306	381	518
512	PORTO ALEGRE	24486	9797	32286	0	186	138	166	97	83	0	183	152	30	30
246	CABO FRIO	0	0	0	0	0	0	0	0	0	0	0	0	0	0
123	SÃO JOSE DOS CAMPOS	54	480	2	0	3	4	0	130	0	0	0	0	12	0
852	FORTALEZA	191	51	0	0	0	1	9	0	0	0	5	0	12	0
482	FLORIANÓPOLIS	755	504	40	0	0	3	3	5	1	0	19	259	7	0
922	MANAUS	1359	1282	57	0	0	0	1	0	1	0	47	4	440	0
192	CAMPINAS	0	4	8	0	0	0	0	0	0	0	0	0	0	0
244	BARRA DO PIRAI	0	0	0	0	0	0	0	0	0	0	0	0	0	0
612	BRASÍLIA	1126	343	1	0	0	3	5	18	0	0	0	0	10	0
473	BLUMENAU	767	780	20	0	0	4	14	5	0	0	34	85	0	0
912	BELEM	740	209	15	0	0	0	0	19	0	0	275	0	0	0
152	SOROCABA	3	0	0	0	0	0	0	0	0	0	0	0	0	0
732A	EUNAPOLIS	0	0	0	0	0	0	0	0	0	0	0	0	0	0
383	SALINAS	0	0	0	0	0	0	0	0	0	0	0	0	0	0
422	PONTA GROSSA	0	0	0	0	0	0	0	0	0	0	0	0	0	0
242	PETRÓPOLIS	0	0	0	0	0	0	0	0	0	0	0	0	0	0
312	BELO HORIZONTE	4261	1369	8	4	0	0	0	22	3	0	288	0	0	0
712	SALVADOR	65	10	0	0	0	0	0	0	0	0	8	0	0	0
125	GUARATINGUETÁ	1	0	0	0	0	0	0	0	0	0	0	0	0	0
122	TAUBATE	5	0	0	0	0	0	0	0	0	0	0	0	0	0
243	VOLTA REDONDA	1	0	0	0	0	0	0	0	0	0	0	0	0	0
194	PIRACICABA	0	0	0	0	0	0	0	0	0	0	0	0	0	0
982	SÃO LUIS	1228	376	185	0	0	26	0	0	0	0	1	1	30	0
186	RIBEIRÃO PRETO	45	229	8	0	0	2	0	6	0	0	4	0	0	0
822	MACEIÓ	51	69	275	0	0	0	47	7	0	0	0	0	0	0
412	CURITIBA	10250	4356	1921	0	22	598	164	90	82	0	197	1019	83	14
326	ALÉM PARAÍBA	0	0	0	0	0	0	0	0	0	0	0	0	0	0
322	JUIZ DE FORA	1	0	0	0	0	0	0	0	0	0	0	0	0	0
812	RECIFE	80	10	0	0	0	14	0	15	0	0	4	0	0	0
382	MONTES CLAROS	0	0	0	0	0	0	0	0	0	0	0	0	0	0
792	ARACAJU	48	116	0	0	0	0	0	0	0	0	0	0	0	0
272	VITORIA	308	133	1	0	0	0	0	0	0	0	0	0	0	0
642	CAXIAS DO SUL	4	4	1	0	0	0	2	0	0	0	0	0	0	0
168	SÃO JOAQUIM DA BARRA	14	0	0	0	0	0	0	0	0	0	0	0	0	0
455	FOZ DO IGUAÇU	80	8	0	0	0	0	1	36	0	0	0	0	0	0
474	JOINVILLE	383	350	28	0	0	0	27	0	0	0	5	56	0	0
332	GOVERNADOR VALADARES	0	0	0	0	0	0	0	0	0	0	0	0	0	0
446	LOANDA	0	0	0	0	0	0	0	0	0	0	0	0	0	0
632	ARAGUAINA	456	149	8	0	0	0	4	0	0	0	125	0	71	0
442	MARINGA	34	0	52	0	0	0	1	64	0	0	0	0	0	0
452	CASCAVEL	0	0	0	0	0	0	0	0	0	0	0	0	0	0
842	NATAL	374	38	1	0	0	1	0	0	0	0	1	0	0	0
824	MORRINHOS	0	0	0	0	0	0	0	0	0	0	0	0	0	0
172	SÃO JOSE DO RIO PRETO	0	0	0	0	0	0	0	0	0	0	0	0	0	0
247	CAMPOS DOS GOYTACAZES	0	0	0	0	0	0	0	0	0	0	0	0	0	0
655	SINOP	0	0	0	0	0	0	0	0	0	0	0	0	0	0
928	BOA VISTA	34	175	12	0	0	0	0	76	0	0	7	0	46	0

Based on the flows (traffic matrix), identify the necessary capacity of each connection for each combination of flows over physical routes. There are two types of algorithms to solve this kind of problem: algorithms using continuous attribution or discrete attribution—Serial Merge.

4.3.4 The process governing the elimination of potential nodes

The elimination of potential nodes, when the associated volume is smaller than a given defined volume, isn't direct. You eliminate the potential nodes one by one. It has to be done in this way, because when you eliminate a node, the users associated with it will be reassociated to other nodes, which may or may not already have the minimum necessary volume. Some nodes, whose volume originally wasn't enough after the elimination of a node, may now become eligible.

Therefore, it is important to have proper node-elimination criteria. Typically, we follow criteria, such as:

- Nodes with the smaller number of *traffic units* go first
- Nodes with smaller traffic volume go first (in or out)
- Nodes near to a survival node go first

It is important to understand, however, that several other criteria may be used, including arbitrary criteria linked with administrative aspects.

4.3.5   Clustering curve

When proceeding with the topologies identification, you also identify what we call the "clustering curve." The clustering curve is an interesting partial result, which shows us the correlation between percentage of users per distance range when we vary the number of aggregation nodes (topological scenarios). This curve gives a clear view about the call-center users' geographical dispersion. Usually, we adopt nine distance ranges:

Range number	Distance
0	X<=20km
1	20km<x<=50km
2	50km<x<=100km
3	100km<x<200km
4	200km<x<=300km
5	300km<x<500km
6	500km<x<=700km
7	700km<x<=1000km
8	x>1000km

Through this curve, you can clearly see how the percentage of users (callers/receivers) per distance range vary when you vary the number of nodes (topology variation).

This is important information, because it shows clearly how the increase or decrease of the number of nodes (topological scenarios) is influential in the call-center traffic distribution.

In addition, this information helps to identify how far from ideal a structure may be without the need for a complete study.

Example:

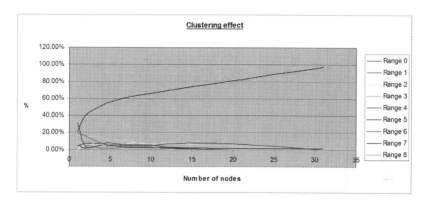

As we can clearly see in this particular example the Δ of variance of percent of users in the closer ranges stops increasing at the same rate around seven nodes.

Number of nodes	% Range 0 (local)	Δ of variation of the number of nodes	Δ of variation of the number of workstations	Δ Per node
1	21.10%			
2	39.78%	1	18.68%	18.68%
4	51.54%	2	11.76%	5.88%
5	56.18%	1	4.64%	4.64%
7	61.07%	2	4.89%	2.45%
8	62.83%	1	1.76%	1.76%
10	66.25%	2	3.42%	1.71%
12	69.38%	2	3.13%	1.57%
20	81.69%	8	12.31%	1.54%
31	97.17%	11	15.48%	1.41%

4.3.6 Topological scenarios generation

The results of this phase are all possible topologies, including the number and location of the aggregation nodes. Here you have all results of all elimination cycles (scenarios)—understanding scenarios as the minimum amount of traffic that makes a node feasible.

4.3.7 Testing the alternatives

Initially, it is important to understand that what is really going to define the ideal design is the interconnection costs (access and backbone).

We know that, in general, the access costs tend to be inversely proportional to the number of aggregation nodes. This means, in general, that the more nodes we have, the cheaper the access tends to be. On the other hand, the more nodes we have, the more expensive the backbone links and the backbone hardware tend to be.

Here it is also important to understand that those tendencies are not absolute and there are tariff systems in which distance isn't a relevant cost definition factor. In these environments, topology may not be as important, and the statements made before may not hold true. From the analytical process point of view, a tariff systems in which distance isn't a cost definition factor doesn't make any difference. In these environments, what usually happens is that we tend to identify the best alternative as a star topology with the centers of the star being the attendance sites.

In this phase, we identify the combination of backbone, access, and hardware, which produces the lower costs. This analysis is executed based on the possible topological scenarios.

This process consists of calculating the costs to provide connectivity for each topological scenario, using the available alternatives (service providers and technologies).

Once we calculate such costs, we select the alternatives (access, backbone, and hardware) capable of generating the lower costs.

Our challenge is to identify which combinations (topologies, interconnection costs, and hardware costs) produce the cheapest designs. The identification of these designs usually happens through exhaustive search (verifying all possible scenarios). This calculation follows the sequence for each scenario:

- Backbone-cost calculation
- Access-cost calculation
- Hardware-cost calculation

This sequence demands some preliminary calculations:

- Nodes traffic volumes (based on the number of users associated with each node)
- Traffic flows—traffic matrixes (based on the volume in minutes)

In other words, this is what happens:

- Calculate the ideal backbone for this scenario (combination of flows, physical routes, and services)

  a) Aggregation nodes traffic-load calculation
  b) Backbone-flows identification
  c) Identification of the available interconnections
  d) Identification of the physical routes available

- Calculate the ideal access for the scenario
- Calculate the ideal hardware for the scenario

Summarizing, for each topology we identify the ideal combination of flows flowing over the physical paths connecting the backbone nodes; we calculate the ideal backbone cost (with all services alternatives available); and we calculate the hardware and access cost. In the end, we have an ideal structure to minimize the transport costs in a specific topological scenario. By repeating the process for all scenarios, we identify the ones with smaller costs. In the next topic we are going to discuss each step with details.

4.3.8   How to select the best alternatives

As explained before, we analyze each topology generated by calculating the cost of the backbone, the access, and the hardware. If our intention is to build a business case, we can also vary the traffic volumes. The topology-cost calculation encompasses four phases:

- Backbone calculation
- Access calculation
- Hardware calculation
- Calculation of the total costs

4.3.8.1 Calculating the backbone cost

Initially, we identify the flows among the aggregation nodes (traffic interest). This is based on the data and voice-traffic matrixes of the traffic units (consolidated by nodes), assuming that the services generate traffic in/out, which is going to flow between the users and the aggregation nodes and between the nodes themselves. It is valid to mention that at this point the availability or lack thereof of physical interconnections to support the flows is not an issue.

These flows will be combined over possible physical routes in order to identify the ideal backbone (ideal in economical terms—let's be clear).

Once we have the flows, we are going to use abrangent trees determination with minimum restriction techniques—MST. The algorithms to determine such abrangent trees have as their objective the interconnection of the backbone nodes, trying to minimize the number of connections (path length). The tree is called "abrangent" because it includes all aggregation nodes (to a given topology).

In the MST logic there isn't a limit to the number of nodes supported by the tree branches. As an example of algorithms belonging to this class we could mention Prim and Kruskal. To calculate properly a backbone in each topological scenario it is necessary to follow these steps:

- For each topology, we analyze all possible physical routes
- For each possible combination of routes, we analyze all possible combinations of applications flows
- For each possible combination of flows over routes, we test all possible service providers and all possible technologies available in each interconnection
- Doing this, we find the optimum backbone cost

In addition, as a byproduct, we execute all calculations and generate all the configuration logic of the equipment.

Example:

Just to illustrate the complexity of the problem, let's assume a situation in which we locked the topology into seven nodes, assuming a call center with two attendance sites and six services. We are also assuming that we are going to compare five service providers, and each one has six alternatives of services. In this specific topology we have:

- $6 \times 6 \times 6 \times 6 \times 6 \times 6$ possible physical routes = 46,656
- Six services and two attendance sites
- Five service providers with six alternatives of interconnection services each
- Number total of alternatives compared for this topology: 46,656 $\times 6 \times 5 \times 6 = 8,398,080$

Graphically this situation can be represented as follows:

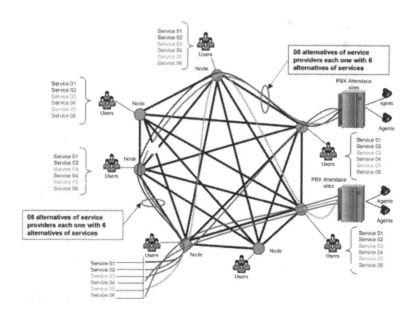

The number of alternatives analyzed grows enormously when the number of nodes grows:

$4^4 \times 6 \times 5 \times 6 = 46{,}080$
$5^5 \times 6 \times 5 \times 6 = 562{,}500$
$6^6 \times 6 \times 5 \times 6 = 8{,}398{,}080$
$7^7 \times 6 \times 5 \times 6 = 148{,}237{,}740$
$8^8 \times 6 \times 5 \times 6 = 3{,}019{,}898{,}880$
$9^9 \times 6 \times 5 \times 6 = 69{,}735{,}688{,}020$
$10^{10} \times 6 \times 5 \times 6 = 1{,}800{,}000{,}000{,}000$

Other interesting aspects to notice here are:

- What would be the possibility of identifying the ideal design if analyzing a backbone with ten nodes manually?
- What is the chance that the manually identified design be even among the best 10 percent?

This example shows us why we need tools to help us design capitation structures properly and why designs made through these tools tend to be so much better (and generate substantial savings).

If we take the example of five service providers, each one offering six different services, and each one of these services maybe using different pricing strategies and demanding different parameters to have their prices calculated (prices, total volume, bandwidth, distance, taxes, state boundaries, countries boundaries, granularity, and minimum traffic committed, just to mention some of them), then we can clearly understand how difficult it would be to compare the alternatives.

4.3.8.2 Calculating the physical connections

At this point, we calculate all possible physical connections among the aggregation nodes and all possible routes. Here it becomes necessary to introduce constraints to limit the excessive number of physical connections.

If no constraints are defined, the calculations will assume that all connections are feasible. A situation like that is not recommended, because it generates an excessive number of possible routes, overloading the computer executing the processing without adding value.

As already mentioned, the lack of limitations in terms of physical connections is very rare.

Example:

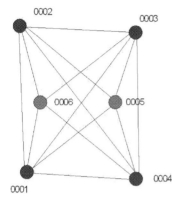

As can be seen in the picture on the left, each node theoretically can be connected with all others, and all can be connected to the attendance sites (black). Without any constraint, we will generate a spreadsheet with columns "from" and "to" describing all possible connections and its distances.

The backbone depicted in the previous picture would have fourteen physical connections.

4.3.8.3 Calculating the routes

In this phase, we calculate all possible routes interconnecting the aggregation nodes. Here also there is the need for limiting the number of hops to avoid an excessive number of possible routes. This means that we are going to use algorithms—Minimum Spanning Tree with Hop Restrictions.

The algorithms Esau-Williams and Sharma are two examples of Minimum Spanning Tree with Hop Restrictions.

The algorithms to determine the minimum path have as their objective, given a predefined group of nodes, to identify the shortest path. We also can include in this class the Bellman-Ford and Dijkstra algorithms.

Considering a capitation structure with a defined topology and with the traffic flows already known, let's assume that we want to identify the ideal voice flow and the ideal structure.

This is an routing-optimization problem. The flow-deviation algorithms and the algorithm Bertsekas-Gallager are two possible methods that can be used to optimize this routing. This analytical process occurs in the following way:

Example:

Follow the previous example, but introduce some limitations in order to reduce the number of available physical connections.

The available physical connections are:

Number	Connection	Number	Connection
1	0001 0002	7	0002 0006
2	0001 0004	8	0003 0004
3	0001 0005	9	0003 0006
4	0001 0006	10	0004 0006
5	0002 0003	11	0004 0005
6	0002 0005		

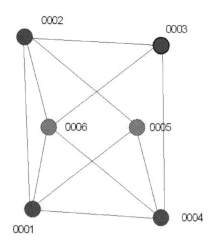

Based on the physical connections, we identify all possible routes linking each aggregation node to the attendance sites. In this example, our traffic interest is concentratd into only two attendance sites (005 and 006), and we are not allowing flow thorough these two nodes. These are the routes available for a flow from node 0001 to the node 0006:

Route	Sequency
1	0001 0006
2	0001 0002 0006
3	0001 0002 0003 0006
4	0001 0002 0003 0004 0006
5	0001 0004 0006
6	0001 0004 0003 0006
7	0001 0004 0003 0002 0006

Generating these routes for every node, we identify all possible routes. Here we should introduce constraints.

Example: No route can have more than three hops. In our example, if we adopt these criteria, the routes 4 and 7 would be discharged as alternatives. Using this criteria, we would have just five routes linking node 0001 to node 0006.

4.3.8.4 Identifying the ideal backbone for the topology

Now that we have identified the routes and the flows, we have to proceed to search to find which combination of flows over physical routes produce the least costly backbone. Following the previous example:

	Route	Sequence
Node 1	1	0001 0006
	2	0001 0002 0006
	3	0001 0002 0003 0006
	4	0001 0002 0003 0004 0006
	5	0001 0004 0006
	6	0001 0004 0003 0006
	7	0001 0004 0003 0002 0006
	Route	**Sequence**
Node 2	1	0002 0006
	2	0002 0001 0006
	3	0002 0001 0004 0006
	4	0002 0001 0004 0003 0006
	5	0002 0003 0006
	6	0002 0003 0004 0006
	7	0002 0003 0004 0001 0006
	Route	**Sequence**
Node 3	1	0003 0006
	2	0003 0002 0006
	3	0003 0002 0001 0006
	4	0003 0002 0001 0004 0006
	5	0003 0004 0006
	6	0003 0004 0001 0006
	7	0003 0004 0001 0002 0006
	Route	**Sequence**
Node 4	1	0004 0006
	2	0004 0001 0006
	3	0004 0001 0002 0006
	4	0004 0001 0002 0003 0006
	5	0004 0003 0006
	6	0004 0003 0002 0006
	7	0004 0003 0002 0001 0006

In this example, we have the flows moving to and from only to two nodes (005 and 006—in the previous picture we show only the routes going to node 6). There isn't any traffic interest among the nodes 0001, 0002, 0003, and 0004. As already mentioned, we test all possible combinations of flows over the available physical routes. When doing so, we identify the demand (bandwidth) of each physical connection. Based on this demand, we calculate the cost if the link was contracted from every available type of service from every available service provider.

We usually have the possibility of setting the tool to instead of calculating the ideal technology, service provider, or vendor in a link-by-link or node-by-node basis, to do it analyzing the alternatives as a whole. For instance, instead of showing which service provider is more cost effective for each specific circuit, we would calculate which service provider can offer the cheaper solution considering all connections. In the same way. instead of showing which vendor has the best price for each node configuration, we can analyze which vendor can provide the more cost-effective proposal, assuming that it would provide the services to all nodes, as can be seen, these kind of detailed verification would be virtually impossible manually. Therefore, we can execute the alternative selection using two possible selecting strategies:

- Per connection: selecting the lower cost of each connection, considering all available services and providers. Connection-by-connection basis.
- Per provider: selecting the lower cost, considering all connections contracted to only one provider. Provider-by-provider basis.

4.3.8.5 Calculating the access cost

If we know the caption node's location and the users associated with the nodes, it is easy to calculate the dialed-access costs (part going through the public network regardless of the direction—in/outbound). The access costs also are identified by the interconnection-cost calculators.

The access cost is the cost to connect the user with its nearest capitation node. This cost can be supported either by the users or by the call-center operator.

Usually, in North America the companies have contracts with the ILECs under which they pay flat rates per spoken minute, regardless of the user location. However, depending on where in the world you are, situations where the users pay for the calls are also common.

The general trend, mainly in countries where the local calls are free of charge, is to have both local and 1-800 numbers. For instance, if a user is originating a call within an area code where the company has a local number, the 1-800 number wouldn't work. If the users call the 1-800 number within these areas, they will hear a message that gives the correct local number to call for this service.

Companies that adopt this strategy usually have the same basic number in all areas, changing only the prefix or area code. For instance, the number in Toronto would be (416) 928-1788 and in Ottawa (613) 928-1788. This arrangement is extremely convenient since the user will only have to remember the same number for any geographic region.

Once we have identified the best configuration possible for a defined number of nodes, we run the process again, changing the minimum number of users per node (which may or may not imply changing the number of nodes). We do this successively, until we verify all possible alternatives and isolate the most cost-effective ones. Here, we calculate the smaller access cost for each topology. In the same way as when calculating the backbone, we may use two different strategies:

- Per connection: selecting the lower cost of each connection, considering all available services and providers. Connection-by-connection basis.
- Per provider: selecting the lower cost, considering all connections contracted to only one provider. Provider-by-provider basis.

These calculations are executed by a family of software called "interconnection-cost calculators" based on the parameters defining the costs.

4.3.8.6 Calculating the hardware cost

Here, we calculate the cost of the hardware associated with each node for each topology. Through this process we identify the cost of each type of hardware available (being considered as alternatives) based on the traffic volume associated with the node. This volume usually is defined in minutes (even if we set indirect parameters such as number of users or calls).

We built the "hardware calculators" based on the hardware of the typical node defined and on the alternatives of the hardware providers being considered. As already mentioned, we must be able to establish the correlation between traffic volume and total cost of ownership of hardware on a monthly basis.

4.3.8.7 Calculation of the total costs and selection of the best design

Once we have executed all calculations (backbone, access, and hardware) for each topological scenario, we are able to identify the ones that produce the smaller total cost. Here it is important to keep in mind that this problem is a heuristic problem and therefore may encompass more than one possible solution.

The outcome of this phase are the more cost-effective designs (note that not only one), including all details necessary to really implement them. Each alternative design must encompass the following information.

Backbone:

- All routes interconnecting each aggregation node
- All flows among all aggregation nodes
- Indication of all links (physical connections) used (point A, Point B)
- The bandwidth of each link used
- The ideal technology to be used by each link
- The ideal service provider for each link
- The price of each link
- The total backbone cost

Access:

- Indication of all access used by the callers/receivers of the calls (End A, End B)
- Each node's number of trunks per service provider
- Each node's access traffic volume and cost per type (long distance, local, mobile)
- Total access cost

Hardware:

- Each node's hardware cost
- Each node's hardware vendor
- Total hardware cost

General information:

- Total structure cost
- Number of users, calls, or minutes, which makes the node feasible
- Number of nodes

With this information, we have all the necessary elements to really implement the structure. Of course, we may make adjustments, and negotiable aspects may play their role also. For instance, if just a handful of trunks were recommended to be contracted from a specific service provider, it would make sense from the operational point of view to contract all of them from the main provider.

4.4   Final selection of designs and adjustments—Phase 3

Using the best theoretical designs, we have now to match them against the technical, administrative, and negotiation realities of the organization. Proceeding in this way, we refine our search and select the really good ones.

Initially, we have to verify the differences between the scenarios and the organization's realities. The objective is to verify which design (or designs) would demand less effort and less time to be implemented.

We also have to compare the designs to the contractual context of the organization by verifying which design would cost less in contract rearrangements—avoiding penalties for instance, which are relevant cost factors.

We also have to verify compatibility with the existing hardware (which may be owned by the organization) with the proposed designs. Although we usually consider the hardware costs when preparing the designs, if the organization already has its own hardware, it eliminates the need for this cost and therefore using the existing hardware has to be considered.

It is also important to keep in mind that having the possible designs we will be particularly well equipped to negotiate with the service providers, and the parameters used to prepare the designs may change due to these negotiations. (It may be a back and forward process—see cases.)

For instance, if just a small number of the circuits were recommended to be contracted with a given service provider by the design generated by the design tool, it may make sense from the operational point of view to contract all circuits from the same provider, so long as we manage to negotiate the same prices as from the cheapest alternative.

In addition, having several design alternatives makes it possible to confront the service providers not only with lower priced alternatives but with different transport strategies. For example, if the value of spoken minutes in the long distance calls were not below X, you may use your private network where the cost is 0.8X. (We discussed in more detail the several aspects involved in a negotiation in chapter 2). At this point, we also have to decide about our backup strategy.

In summary, we are going to make adjustments by contemplating technical, commercial, contractual, and strategic issues.

This phase could be divided into the following sub-phases:

- Understanding the actual situation
- Comparing the proposed designs in the contractual context
- Comparing the proposed designs in the technical contexts
- Using the designs to negotiate
- Defining the backup strategy
- Calculating ROI
- Adjusting the chosen design

4.4.1   Understanding the current situation

Having a clear understanding of the current situation of the telecommunications infrastructure makes the team in charge of the project able to define which designs are really ideal in terms of cost, time, and implementation effort. In addition, having a clear view of the current situation makes it possible to compare the foreseen costs with the current ones. Having this information, you will be able to see the project's return over investment (ROI).

This kind of analysis/redesign normally implies substantive savings. When executing this process, you also identify the need for equipment replacement (or upgrade) and the project's time frame. Also, the redesign process helps with the analysis of strategic alternatives, such as outsourcing and external management.

4.4.1.1 Confronting the designs with the contractual reality

Here we have to verify each contract and identify its scope, time to termination, and penalties for early termination. Through this verification we have to understand exactly the cost involved if we terminate each contract completely or if we reduce its scope. We have to be able to see how this cost changes with the time (today, three months, six months, a year). These values are crucial when calculating the ROI because the early-termination penalties usually are quite significant cost factors.

4.4.1.2 Confronting the designs with the technical and operational realities

It is very important to understand how the call-center structure is physically and technically structured because we have to know things such as whether the equipment deployed is bought or rented and if the equipment is old or up-to-date. It is crucial to understand what limitations the hardware being used imposes on rearranging the structure. (Hardware in this context doesn't mean only network hardware.)

In addition, technical understanding is crucial to allow you to see all operational aspects of the organization. It is important because you may see situations such as an optimal design pointing to a topology with nodes in different locations than the operational centers of the organization.

In situations like that, it is very important to match as much as possible the theoretical designs with the operational context of the organization.

If that is not possible, you must consider the cost associated with the operational rearrangements in the cost of each scenario.

4.4.2   Using the designs to negotiate

Another aspect to be considered is that, having the theoretical designs, you can confront the service providers' different transport strategies. Although it may not be a typical procedure, it can be done, and, in some cases, must be done. If you manage to attain some commercial advantage directly (getting a discount from the service provider that makes the implementation of a new structure pointless, for example), you may review the whole strategy.

4.4.3   Defining the backup strategy

When designing a call-center structure you have to define a backup strategy, the definition of which is very important, because it is what allows you to define what we call "reaction capability" of the structure. When designing a call-center structure, you must define what might happen if each one of its components failed. Of course, the level of reliability of a structure depends on the kind of activity each organization and its business side must be able to inform how much the down time costs

This information—how much downtime costs—isn't always easily calculated; however, a honest effort must be made because this cost will be the basis on which several backup alternatives will be considered. Here are some examples of backup strategies:

1) Each aggregation node will have at least two data circuits, one primary and another a backup secondary.
2) The attendance sites and all aggregation nodes will be connected to at least two COs (central office) of each service provider, and arrangements should be made to guarantee that, in case of problems in one of the COs, the traffic will flow entirely though the other.
3) The attendance sites and all aggregation nodes will be connected to at least two power supply substations, and arrangements should

be made to guarantee that, in case of problems, one of them will be able to support the site alone.

4) Each aggregation node will have more than one service provider. In case of fail (or overflow) in one service provider, the calls are automatically forwarded to the other.

Having defined the macro backup strategy, you still have to define aspects such as whether the backups will be kept empty and be activated only in case of failure of the main alternative or if they will be used on a regular basis (with spare capacity).

When designing the backup structure, usually you should remove all connections belonging to the service providers chosen in the main design. By doing that, you force the backup alternative to be provided by a different service provider than the one supporting the main design.

"On-use" backup structures used to be more reliable than the "just-in-need" ones. A resource used only occasionally has a much bigger chance of being defective when you need it. Furthermore, if you are using a resource that would otherwise be empty, you are also saving money and keeping them monitored.

4.4.4   Redesigning the structure

Once you verify the current situation and understand the contractual, technical, and negotiable context and have defined the failure strategy, you should refeed the algorithms with the new inputs. Once you recalculate all the designs, you have the possible scenarios in their final form. They are no longer theoretical designs but practical ones.

4.4.5   Calculating the ROI

Once you define the best designs, you calculate the return over the investment of each one of them, considering not only the monthly operational cost but also the implementation costs, including but not limited to:

- Consulting and specialized services
- Early-termination penalties
- Negotiation costs
- Hardware renovation cost
- Migration costs
- Operational costs
- Management costs
- Internal workforce cost

Once you have all these costs properly mapped for each design, you should choose the one that combines the best alternative for the organization.

Here it is important to mention that we don't see the design tools as a panacea. It doesn't matter how good the tool is, the designs will be as good as the information used to feed the tool, and in all cases the human evaluation is a critical factor. In this sense, we could say that designing call-center structures is as much an art as it is a science. The design tools work as powerful calculators, which help the designers to do all required calculations quickly and precisely (which, without the tools, would be virtually impossible to do), but it is a mistake to think that they produce results by themselves. At the end of the day, it is the person evaluating the tool results and matching them against the contextual realities who will find the adequate design.

4.5 Varying the volumes and building dynamic business models

Although the identification of an ideal structure to support a given traffic volume is in itself very important, the deployment of algorithms allows you to do more. Having tools that allow you to calculate these structures fast opens the possibility to make many calculations, using several traffic volumes. Through these analyses, you can establish the correlation between volume and cost (and eventually the associated revenue).

The possibility to set up many volume scenarios is extremely useful because it allows you to see how the infrastructure cost changes depending on the volumes transported. Therefore, you can produce a graphic correlating these values.

For instance, if you are analyzing a large call center, it becomes possible to make simulations in which you can see how the cost varies when you vary the transported volume. It allows you to identify in advance the impact of a new service in terms of infrastructure costs. It also becomes feasible to verify if a service-provider proposal is better, considering not only the current traffic volume but also a future one.

In the same way, you can identify the correlation of revenues per volumes and in doing so you are able to identify the correlation between traffic volumes versus infrastructure cost and traffic volume versus revenue. With this information it becomes possible to identify:

- The company cash flow
- Breakeven point—the minimum amount of traffic necessary to make a project feasible
- ROI (Return on Investment)
- Profitability per traffic volume

In summary, you are able to build the business models and consequently see the limits of feasibility involved in each business initiative and alternative.

This kind of knowledge is crucial when planning investments, because it allows the investors to evaluate properly the risk involved in the endeavour. It also enables the simulation of different charging strategies, which evaluates the impact on the revenue stream. Having a tool that allows the investors to see the different potential scenarios is extremely helpful when looking for funding.

Having a dynamic model enables you to analyze how the variables involved influence each other and verify how changing each one of them affects the overall cost of the structure.

These correlations allow the design team to make decisions with respect to a wide range of issues, from the purely technical to strategic. For example:

Market strategies

- The minimum amount of users necessary to make the services feasible
- Services provided
- How much is charged by each service

Operational strategies

- Who pays for the access (availability or not of 1-800 access)
- Whether or not the company provides local numbers
- Services are provided only through the IVRs, only by live attendants, or both
- Insource or outsource the live attendant and transport

Technical strategies

- Whether or not to use traffic caption network of nodes
- Distribute IVRs or not
- Voice compression rate
- Acceptable quality of service
- Hardware and interconnection providers
- Interconnection technologies

Decisions such as these are very hard to make without an automated tool that analyzes all aspects of the issue. It becomes even harder when varying the demand that the call center is suppose to handle.

By using tools to perform this analysis, you can establish the correlation between all variables involved. In other words, you can model the problem.

The capacity to analyze the structure dynamically allows you to verify the cost associated with implementing each new service, assuming different percentages of the usage. Therefore, if you have an income or revenue associated with the transactions performed by the call center, we can even—as alluded to above—produce a return-on-investment analysis and show where the breakeven point is as with respect to an individual service.

The analysis provides the ability to make well-educated decisions as to who pays for the access as well as which services will be offered only through the IVRs or only through live attendants.

For instance, if you verify that 90 percent of your users are located inside the area codes of your nodes you may consider the alternative of not providing 1-800 services for these areas at all.

Example:

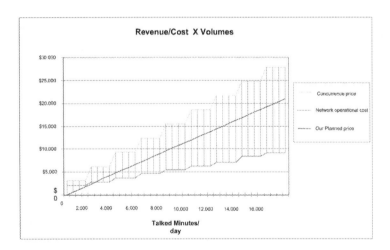

The graphic shows how the infrastructure cost varies when the volume of minutes varies and also shows how the revenue curve behaves when the volume of minutes varies. Through this graphic you can see the breakeven point around three thousand minutes a day. This point is where the revenue curve crosses the operational-costs curve.

4.6 Comparing the results with an infrastucture already in place

If you have an infrastructure already in place (which is usually the case), you should compare the costs of the proposed new structure with the actual one. At this point, it will become clear which cost factors in the actual structure could be reduced, and you can produce a very straightforward, high-level managerial report that compares the actual expenditures with the proposed ones. Typically, you can state as clearly as possible the potential of savings and the investment necessary to implement the proposal, and you can also calculate the savings and return on investment of the project as a whole.

4.7 Analyzing alternatives such as outsourcing

You can further analyze alternatives such as outsourcing or external management. Having a clear view of how much it would cost to build its own infrastructure gives the organization a better understanding of how much would be reasonable to pay for an external management or a completely outsourced solution. In addition, when soliciting bids for an outsourced solution, the process is again simplified since the parameters around the requirements are predefined and constant. This gives the organization the ability to compare apples to apples when reviewing the proposals and so make the evaluation process simple in this respect.

Chapter 5: Planning a Transport Structure

In this chapter, we will discuss an example of a design of a real call-center structure. The idea is to identify the best structure to support an outbound call-center operation, trying to guarantee the minimum transport cost.

5.1 Premises adopted

In this section, we will define the premises adopted in the calculations, trying to identify the best strategy to bring the current costs down (technical and negotiable). It is important to remember that this particular organization is an outbound-only operation.

We will model the operation using typical values for interconnection costs and hardware in the market where the organization operates in order to be able to simulate the possibility of having a private network.

5.2 Interconnection costs

Now, we will show the reference cost for interconnections (switched and dedicated) used to execute the calculations. We are going to use only one price reference for data services (dedicated) and only one for voice services (switched). Theoretically, we could use more than one price reference;

however, only if we were using a design tool, would it be feasible to do the necessary calculations (using a per call/connection basis or as a whole).

5.2.1   Voice services (switched)

We are going to use as reference the typical values of one specific service provider. These values don't pretend to be the best achievable ones in this market and therefore are without doubt within the negotiable possibilities of the organization (we are being conservative in terms of interconnection costs).

Telco 1							
Outbound							
To fix lines					**To mobile lines**		
Local	Range 1	Range 2	Range 3	Range 4	Local	Intrastate	Interstate
USD 0.04	USD 0.08	USD 0.08	USD 0.08	USD 0.08	USD 0.55	USD 0.55	USD 0.55
Inbound							
To fix lines					**To mobile lines**		
Local	Range 1	Range 2	Range 3	Range 4	Local	Intrastate	Interstate
USD 0.04	USD 0.08	USD 0.08	USD 0.08	USD 0.08	USD 0.55	USD 0.55	USD 0.55

The calculation process will encompass also the variation of the cost of the local calls in order to identify the breakeven point from where having a network of nodes to distribute the calls becomes economically attractive. This logic also admits an inverse analysis, which tariff we would need to identify as making building a network of nodes uneconomical.

We simulated local tariffs from USD 0.03 to USD 0.02 (the current value paid by the organization is USD 0.04). These values are perfectly achievable by the organization. Therefore, the values used as reference for the simulations were:

Telco 1								
Outbound								
To fix lines					To mobile lines			
Local	Range 1	Range 2	Range 3	Range 4	Local	Intrastate	Interstate
USD 0.03	USD 0.08	USD 0.08	USD 0.08	USD 0.08	USD 0.55	USD 0.55	USD 0.55
Inbound								
To fix lines					To mobile lines			
Local	Range 1	Range 2	Range 3	Range 4	Local	Intrastate	Interstate
USD 0.03	USD 0.08	USD 0.08	USD 0.08	USD 0.08	USD 0.55	USD 0.55	USD 0.55

Telco 2								
Outbound								
To fix lines					To mobile lines			
Local	Range 1	Range 2	Range 3	Range 4	Local	Intrastate	Interstate
USD 0.02	USD 0.08	USD 0.08	USD 0.08	USD 0.08	USD 0.55	USD 0.55	USD 0.55
Inbound								
To fix lines					To mobile lines			
Local	Range 1	Range 2	Range 3	Range 4	Local	Intrastate	Interstate
USD 0.02	USD 0.08	USD 0.08	USD 0.08	USD 0.08	USD 0.55	USD 0.55	USD 0.55

5.2.2   Data services (dedicated)

EIR	CIR	Last mile	CPE	Gateway	Total
128	64	USD 228.38	USD 161.60	USD 240.00	USD 629.98
256	128	USD 308.77	USD 161.60	USD 240.00	USD 710.37
256	192	USD 389.16	USD 161.60	USD 240.00	USD 790.76
512	256	USD 469.55	USD 161.60	USD 240.00	USD 871.15
512	320	USD 549.94	USD 161.60	USD 240.00	USD 951.54
512	384	USD 630.33	USD 161.60	USD 240.00	USD 1,031.93
512	448	USD 710.72	USD 161.60	USD 240.00	USD 1,112.32
1024	512	USD 791.11	USD 161.60	USD 240.00	USD 1,192.71
1024	576	USD 871.50	USD 161.60	USD 240.00	USD 1,273.10
1024	640	USD 951.89	USD 161.60	USD 240.00	USD 1,353.49
1024	704	USD 1,032.28	USD 161.60	USD 240.00	USD 1,433.88
1024	768	USD 1,112.67	USD 161.60	USD 240.00	USD 1,514.27
1024	832	USD 1,193.06	USD 161.60	USD 240.00	USD 1,594.66
1024	896	USD 1,273.45	USD 161.60	USD 240.00	USD 1,675.05
1024	960	USD 1,353.84	USD 161.60	USD 240.00	USD 1,755.44
2048	1024	USD 1,434.23	USD 161.60	USD 240.00	USD 1,835.83
2048	1088	USD 1,514.62	USD 161.60	USD 240.00	USD 1,916.22
2048	1152	USD 1,595.01	USD 161.60	USD 240.00	USD 1,996.61
2048	1216	USD 1,675.40	USD 161.60	USD 240.00	USD 2,077.00
2048	1280	USD 1,755.79	USD 161.60	USD 240.00	USD 2,157.39
2048	1344	USD 1,836.18	USD 161.60	USD 240.00	USD 2,237.78
2048	1408	USD 1,916.57	USD 161.60	USD 240.00	USD 2,318.17
2048	1472	USD 1,996.96	USD 161.60	USD 240.00	USD 2,398.56
2048	1536	USD 2,077.35	USD 161.60	USD 240.00	USD 2,478.95
2048	1600	USD 2,157.74	USD 161.60	USD 240.00	USD 2,559.34
2048	1664	USD 2,238.13	USD 161.60	USD 240.00	USD 2,639.73
2048	1728	USD 2,318.52	USD 161.60	USD 240.00	USD 2,720.12
2048	1792	USD 2,398.91	USD 161.60	USD 240.00	USD 2,800.51
2048	1856	USD 2,479.30	USD 161.60	USD 240.00	USD 2,880.90
2048	1920	USD 2,559.69	USD 161.60	USD 240.00	USD 2,961.29
2048	1984	USD 2,640.08	USD 161.60	USD 240.00	USD 3,041.68
2048	2048	USD 2,720.47	USD 161.60	USD 240.00	USD 3,122.07

For data services, we are also using typical values for frame-relay services in this marketing. Once again, when defining these references we have to be conservative in order to guarantee that the organization can reach these values in a real negotiation.

5.3   Node hardware costs

Before defining the typical prices for hardware, we have to define what will be our "typical" node in terms of hardware. In our particular example, we are assuming that three types of equipment constitute the typical node:

111

- PBX
- Router
- Switch WAN

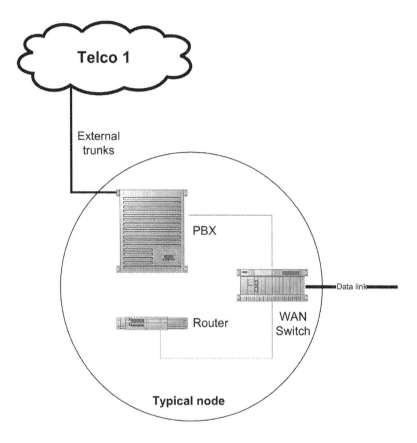

5.3.1 PBX

We assumed a cost of USD 9,000 per card with 30 ports and USD 15,000 per magazine, assuming each magazine supporting a maximum of 10 cards.

We didn't consider the cost of the attendance hardware. In this context, the hardware of the attendant site (let's be clear—associated exclusively with the attendants) will be the same regardless of the topology. Therefore, the term "hardware" here refers exclusively to the equipment that's function is linked with the transport of the calls.

5.3.2 Router

We assumed the cost of the router card with eight ports as USD 600 and USD 3,000 per magazine with a maximum of twenty cards per magazine.

5.3.3 WAN Switch

We assumed a cost of USD 3,000 per WAN switch port and USD 45,000 per magazine with twenty ports.

We assume a TCO (Total Cost of Ownership) of 4 percent monthly including leasing, maintenance, and hardware management for all equipment. All these costs are conservative ones.

5.4 Number and location of the attendance sites

The attendance sites are those where the calls are attended by the human attendants and where you may or may not deploy the IVRs. In this particular case, we had an outbound-only operation, and therefore IVRs are not used.

You could assume two possible distinct scenarios regarding the location of the attendance sites: you could define the location of the attendance sites or you could let a design tool define where it would be best to place the attendance sites.

A design tool could define the best attendance locations based on the number of users located within the defined distance ranges. In this particular case, considering the fact that this is an organization already established and which owns its sites, it was decided to not consider the possibility of relocating the attendance sites. Therefore, we are not analyzing the location of the attendance sites.

The attendance sites are linked with the capitation nodes in such way as to allow the outbound traffic to be transported to the nearest capitation node, using the public infrastructure only between the node and the number called.

The data circuits connecting the attendance sites and the capitation nodes are referred collectively as "backbone."

5.5   Quality of services parameters

We will calculate the number of necessary trunks using the Erlang B formula with an admissible loss of 1 percent.

5.6   Topological analysis (clustering effect)

As already mentioned, in general a capitation node could be defined as a traffic concentration point in a given area code (inbound and outbound traffic) where IVRs may or may not be deployed.

In our example, because it is an outbound operation, the transport cost constitutes the cost of the calls between the site generating the call (the attendance sites) and the users. In an outbound traffic operation with a private transport structure, the call would go through the private network part of the way and get the public network in the other part (going from the nearest node to the user),

Here we are going to demonstrate what we call the "clustering effect" specific for this organization. This effect demonstrates the variation of the percentage of users within each distance range when we vary the number of nodes:

Number of nodes	Local	Between 20 and 50 km	Between 51 and 100 km	Between 101 and 200 km	Between 201 and 300 km	Between 301 and 500 km	Between 501 and 700 km	Between 701 and 1000 km	Above 1000 km
3	44.58%	0.21%	1.31%	6.32%	2.50%	8.04%	5.27%	9.96%	21.81%
6	52.22%	0.56%	2.82%	4.90%	4.60%	11.86%	6.03%	9.63%	7.39%
7	54.36%	0.56%	2.97%	4.98%	4.79%	11.58%	4.61%	9.83%	6.32%
9	58.08%	0.56%	2.97%	5.78%	5.24%	13.56%	5.34%	4.89%	3.59%
12	61.80%	1.62%	3.78%	7.98%	6.30%	11.62%	5.56%	0.60%	0.73%
17	66.48%	2.32%	5.03%	8.78%	3.05%	10.12%	2.89%	0.60%	0.73%
19	68.43%	1.26%	5.46%	8.88%	3.54%	9.19%	1.90%	0.60%	0.73%
31	76.66%	1.70%	6.04%	6.19%	4.52%	2.63%	1.45%	0.60%	0.21%
43	82.71%	1.59%	5.76%	5.21%	2.47%	1.01%	0.47%	0.57%	0.21%
51	85.91%	1.81%	4.57%	4.41%	1.67%	0.97%	0.36%	0.09%	0.21%
54	86.88%	1.90%	4.16%	4.33%	1.13%	0.95%	0.36%	0.09%	0.21%
57	87.76%	1.93%	3.55%	4.05%	1.30%	0.75%	0.36%	0.09%	0.21%
67	90.27%	1.68%	2.29%	3.38%	0.98%	0.75%	0.36%	0.09%	0.21%
75	92.02%	1.25%	2.40%	2.58%	0.51%	0.59%	0.36%	0.09%	0.21%
81	93.13%	1.08%	2.33%	2.15%	0.47%	0.46%	0.30%	0.08%	0.00%
101	95.59%	0.72%	1.80%	1.17%	0.14%	0.43%	0.15%	0.00%	0.00%

We can have a better view through the graphic:

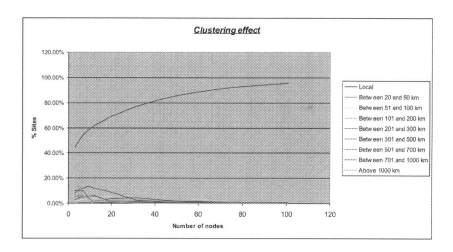

The spreadsheet below gives us a better understanding of the Δ of variation of the percentage of local users when we vary the number of nodes. Note the fact that for this particular organization, the maximum impact in terms of percentage of local users occurs until five nodes, from there the percentage of local users added each time a node is added reduces considerably.

WAN item	01 node (current situation)	03 nodes	06 nodes	09 nodes	12 nodes
Hardware cost	USD 0.00	USD 16,861.50	USD 23,485.50	USD 30,359.25	USD 49,826.25
Backbone cost	USD 0.00	USD 81,667.74	USD 104,188.28	USD 117,312.69	USD 140,411.60
Access cost	USD 827,705.00	USD 782,884.24	USD 755,796.48	USD 735,892.85	USD 701,412.42
Total	USD 827,705.00	USD 881,413.48	USD 883,470.26	USD 883,564.79	USD 891,650.27

This means that, if this specific organization is considering the hypothesis of building its own transport network in order to obtain the benefit of having local call prices for long distance calls, the trend is that these benefits (savings) will be more effective with a topology of five nodes. In the same way, any negotiation for special tariffs in specific locations will tend to be more effective within these five locations.

The calculations necessary to build this graphic were based on the volume of minutes grouped by area codes (the traffic matrix) and, of course, a defined geodesic coordinate for each area code (usually the main city in the area).

5.7   Average call duration

By analyzing the bills we verified that the average call duration of this particular organization is 92 seconds for the success calls (calls where an actual conversation took place) and 30 seconds for the calls without success (short calls). In reality the short calls (measured precisely by the billing system) had an average duration of 12.47 seconds, but because the charging granularity is 30s+6s+6s all short calls are charged as if they had a duration of 30 seconds. In practical terms, all calls with a duration equal to or below 30s are charged as 30s.

5.8   Voice-compression rate

We define the compression rate of the voice channels when flowing through the data network (backbone). The compression-rate definition depends upon the available compression mechanisms and the minimum expected quality of service. Typical rates are 32, 16, and 12K per voice channel. In this particular example we are assuming 16K, which is very typical.

5.9   Simulations and results

Executing the simulations considering topologies with 3, 6, 9, and 12 nodes using the current costs described before (USD 0.04 for local calls and granularity of 30s+6s+6s) the total costs identified were:

WAN item	01 node (current situation)	03 nodes	06 nodes	09 nodes	12 nodes
Hardware cost	USD 0.00	USD 16,861.50	USD 23,485.50	USD 30,359.25	USD 49,826.25
Backbone cost	USD 0.00	USD 81,667.74	USD 104,188.28	USD 117,312.69	USD 140,411.60
Access cost	USD 827,705.00	USD 782,884.24	USD 755,796.48	USD 735,892.85	USD 701,412.42
Total	USD 827,705.00	USD 881,413.48	USD 883,470.26	USD 883,564.79	USD 891,650.27

As you can see, the costs of using topologies with more than one node (current situation) are higher. It happens because the current tariff for local calls of this organization (USD 0.04) is much higher than it could be, because when negotiating the contract, the negotiator knew that the most traffic was long distance and therefore pushed to reduce the cost of this specific type of call, even if the organization had to pay a bit more for local calls. Now let's see what would happen if we had a minute of local calls reduced to USD 0.03 and to USD 0.02 and keeping the topology as a star:

Structure	Hardware cost	Backbone Cost	Access fix-to-fix	Total
Current structure with one node (SP) tariff fix-to-fix USD 0.04	USD 0.00	USD 0.00	USD 827,705.00	USD 827,705.00
Current structure with one node (SP) tariff fix-to-fix USD 0.03	USD 0.00	USD 0.00	USD 783,733.17	USD 783,733.17
Current structure with one node (SP) tariff fix-to-fix USD 0.02	USD 0.00	USD 0.00	USD 739,761.34	USD 739,761.34

As you can see, best tariffs for local calls without changing the topology reduced the overall cost only 5 percent and 10 percent when we applied the tariffs of USD 0.03 and USD 0.02 respectively (only 21.25 percent of the total costs refers to local calls).

Now let's see what happens when we increase the number of nodes to five and transform 50.11 percent of the traffic to local. When we do that, the reduction of the rate of the local calls has a much bigger impact (50.11 percent instead of 21.25 percent).

Structure	Hardware cost	Backbone Cost	Access fix-to-fix	Total
Structure with 5 nodes and tariff fix-to-fix USD 0.03	USD 23,485.50	USD 31,218.90	USD 724,014.26	USD 778,718.66
Structure with 5 nodes and tariff fix-to-fix USD 0.02	USD 23,485.50	USD 31,218.90	USD 620,323.51	USD 675,027.91

In this particular example, creating capitation nodes makes sense if we get a local call around USD 0.02 (considering the price of the hardware, backbone, and access). With five capitation nodes, we manage to bring the costs down from USD 739,761.34 to USD 675,027.91, which is an additional reduction of 10 percent.

Through this example, it becomes clear that the feasibility of a private network to transport the calls has a direct correlation with the difference between the costs of the local and the long distance calls. On the other hand, the cost of the private network varies with the number of nodes deployed, as in this particular example:

Number of nodes	Backbone	Hardware	WAN Cost (without access)	Cost per node
03 nodes	USD 18,731.34	USD 16,861.50	USD 35,592.84	USD 11,864.28
05 nodes	USD 31,218.90	USD 23,485.50	USD 54,704.40	USD 10,940.88
09 nodes	USD 62,437.80	USD 30,359.25	USD 92,797.05	USD 10,310.78
12 nodes	USD 93,656.70	USD 49,826.25	USD 143,482.95	USD 11,956.91

Therefore, the savings through the deployment of a private-transport structure occurs in locations where the difference between the cost of local and long distance calls, multiplied by the volume, exceeds the costs of building a node (the cost per node is indicated in the spreadsheet—hardware and backbone).

Following this logic, in this particular example, if we had a USD 0.02 per minute local and USD 0.08 per minute long distance, we would save USD 0.06 for each minute transported through the private structure (previously long distance transformed to local). This value, multiplied by the number of minutes, is the raw savings of using a private structure to transport the calls. From this value we have to subtract the value of building the node (hardware) and the data link to connect it to the attendance sites (backbone).

Number of nodes	Cost per node	Minimum number of minutes to make a node feasible
Between 01 and 03 nodes	USD 11,864.28	237,285.60
Between 04 and 05 nodes	USD 10,940.88	218,817.60
Between 06 and 09 nodes	USD 10,310.78	206,215.67
Between 10 and 12 nodes	USD 11,956.91	239,138.25

As we can see, for this particular organization within a total universe of 9,000,000 minutes, we have 31.57 percent of the calls local to São Paulo Area (011). Therefore, we have 6,158,000 minutes long distance, which in theory could be converted to local, if we had a private-transport structure. We say "in theory" because a private network wouldn't cover all area codes but just the ones with a reasonable volume of minutes. In this particular example, even if we remove from our calculations the fix-to-mobile calls we still would have 5,850,000 minutes long distance. Therefore, if we calculate the traffic interest for this specific organization for the five main areas of traffic concentration (four more nodes besides São Paulo), we would have 1,138,000 minutes transformed from long distance to local:

Node location	Area code	Cost of the calls destinated to fix-to-fix	Minutes fix-to-fix	Cost per minute today	Cost per minute if local	Value fix-to-fix long distance	Value fix-to-fix local	Difference	Cost per node	Gains for using a private network
SAO PAULO	011	USD 226,621.36		USD 0.02	USD 0.02	USD 0.00	USD 0.00	USD 0.00		USD 0.00
RIO DE JANEIRO	021	USD 110,549.20	495,768.12	USD 0.08	USD 0.02	USD 39,661.53	USD 9,915.38	USD 29,746.15	USD 11,864.28	USD 17,881.87
BELO HORIZONTE	312	USD 50,097.65	225,728.42	USD 0.08	USD 0.02	USD 18,058.27	USD 4,514.57	USD 13,543.71	USD 11,864.28	USD 1,679.43
CAMPINAS	192	USD 38,906.67	215,141.41	USD 0.08	USD 0.02	USD 17,211.31	USD 4,302.83	USD 12,908.48	USD 10,940.88	USD 1,967.60
SALVADOR	712	USD 29,179.26	201,587.01	USD 0.08	USD 0.02	USD 16,126.96	USD 4,031.74	USD 12,095.22	USD 10,940.88	USD 1,154.34

As we can see, assuming the cost of USD 11,864.28 per node (demonstrated in the previous spreadsheet) and the difference between local and long-distance minutes of USD 0.06 (difference between USD 0.02 and 0.08), we would have four additional areas eligible to become nodes. These nodes would make feasible a reduction of USD 22,683.24 (summing of the column "Gains for using a private network") monthly in the current expenditures. In other words, by negotiating better local call prices and building four nodes, we would manage to bring the costs down from USD 827,705 to USD 675,027 (19 percent by just negotiating better one specific type of tariff and building four capitation nodes).

Another interesting consideration is that the cost of building a private-transport structure doesn't grow at the same rate as the volume.

Therefore, when the traffic grows, the chance of more nodes becoming economically feasible also grows.

This example gives us a general understanding of the logic of the process and how we can use this analysis to find our way through the several possible paths to archive-cost reductions.

Chapter 6: Cases

6.1 Call-center operator reevaluates its telecommunications infrastructure

In this example, we will show a case where routing the calls determined the ability to reduce operational costs. This case took place in an outbound call-center operator with approximately three thousand attendants distributed into two sites.

The organization deployed PBXs AVAYA and a wide variety of dialer devices. The average monthly volume was around 8.9 million minutes.

The organization was looking for alternatives to reduce its current telecommunications costs, which were around USD 1.3 million a month.

The first step was to analyze the traffic carefully, identifying the volumes, interest, and profile. The volumes were distributed by service provider as follows:

Provider	Type of service	Minutes	Value
Telco 1	Outbound fix-to-fix local	1,866,772.50	USD 82,320.54
	Outbound fix-to-fix long distance	4,371,791.00	USD 406,827.15
	Outbound fix-to-mobile local	176,224.10	USD 191,909.97
	Outbound fix-to-mobile long distance	85,479.40	USD 68,608.34
Telco 2	Inbound fix-to-fix local	34,917.00	USD 24,492.10
	Inbound fix-to-fix long distance	79,353.40	USD 64,727.37
	Intbound fix-to-mobile local	458.90	USD 110.60
	Inbound fix-to-mobile long distance	378.20	USD 93.22
	Outbound fix-to-fix local	36,124.30	USD 3,109.81
	Outbound fix-to-fix long distance	144,466.40	USD 46,672.05
Telco 3	Outbound fix-to-fix local	101,444.60	USD 21,369.12
Telco 4	Outbound fix-to-mobile local	23,528.40	USD 12,032.28
	Outbound fix-to-mobile long distance	102,727.00	USD 92,524.27
Telco 5	Outbound fix-to-fix local	552,242.69	USD 33,723.17
	Outbound fix-to-fix long distance	1,293,296.12	USD 166,659.53
	Outbound fix-to-mobile local	52,131.94	USD 78,617.23
	Outbound fix-to-mobile long distance	25,287.16	USD 28,105.87
Telco 6	Outbound fix-to-mobile long distance	96,785.39	USD 31,267.98
Total		8,946,623.11	USD 1,353,170.61

The analysis identified the possibility of reducing the current telecommunications costs by 51 **percent** from the current USD 1,353,000. to USD 668,000 per month.

The analytical process demanded the consolidation of all call logs and the recalculation of the calls considering prices if executed by each one of the available service providers. Once this calculation was executed, it became possible to identify the maximum potential savings if each call was executed through the service provider with the lower cost. In addition the volumes per area code and per service provider (calls designated to each provider) were calculated. Such information empowered us to negotiate tariffs.

In practical terms, we managed to bring the costs down as shown in the spreadsheet:

Demonstrative of monthly expenditures with telecom

Ref	Telco 1	Telco 2	Telco 3	Telco 4	Telco 5	Telco 6	Total	Savings
December (baseline)	USD 749,666.00	USD 139,205.00	USD 21,369.12	USD 104,556.55	USD 307,105.21	USD 31,267.98	USD 1,353,169.86	
January	USD 382,244.76	USD 140,696.54	USD 12,930.45	USD 117,507.50	USD 224,345.00	USD 38,081.94	USD 915,806.19	USD 437,363.67
February	USD 450,826.05	USD 118,263.92	USD 56,687.32	USD 135,858.19	USD 245,275.62	USD 7,194.66	USD 1,013,925.76	USD 339,244.10
March	USD 368,443.25	USD 81,933.52	USD 117,815.41	USD 162,800.36	USD 220,228.63	USD 2,387.48	USD 953,588.65	USD 399,581.21
April	USD 377,397.33	USD 63,408.74	USD 113,495.03	USD 99,673.48	USD 30,205.92	USD 685.10	USD 685,069.60	USD 668,100.26

The actions, which led to these savings, identified through the analytical process, were:

1) Least-Cost Routing: guaranteeing that each call was executed using the provider whose price was the cheapest. (To be able to do that we have to do some rearrangements in the infrastructure.)

2) Deployment of GSM gateways: through these devices, we managed to substantially reduce the costs of calls destined to mobile phones.

3) Re-negotiation of charging granularity: the change of the charging granularity from 30s+6s+6s to 6s+6s+6s generated significant reduction of the costs of short calls—calls whose average real duration was less than 15s.

4) Tariff re-negotiation: through well-informed negotiations (traffic mapped in terms of volume, interest, and service provider termination) it becomes possible to attain substantial reductions in the tariffs in the areas where the volume of calls was bigger.

The data-gathering process followed the methodology described in this book.

The evaluation process demanded the consolidation of all call logs (twenty-eight invoices) and the generation of the traffic matrix (per area code). The process as a whole (data gathering and analysis) took one month and the implementation four months. The necessary investment to re-arrange the infrastructure demanded USD 350,000.

6.2 Brazilian Financial Company optimizes its call-center transport structure

This case describes a project in which worldwide, diversified, financial-services company analyzed its telecommunications infrastructure in its inbound call-center operations in Brazil. It illustrates how a good analysis can be the determinant when negotiating with the service providers. The basic facts were:

- The total of spoken minutes per month was 4,519,676.
- The number of calls was 1,396,327.
- The average call duration was 3.23 minutes.
- Calls were originated from all parts of Brazil and handled by an attendance site located in Rio de Janeiro.
- It had a contract with one of Brazil's main ILECs, paying a flat rate per minute.

The company was studying alternatives to reduce its current telecommunications costs of approximately USD 828,000 per month.

Studying the traffic, potential savings of approximately USD 401,000 per month were identified, bringing the current cost down by 48 percent. The redesigned structure would cost USD 558,000 per month.

The alternative service providers available were the four main Brazilian telcos, all of them providing either dedicated or switched connections. In Brazil, the calls within the same area code are charged and the 1-800 calls can be charged differently depending on where they originated.

The call distribution is shown below. As can be seen in the HCF (Hourly Concentration Factor) graphic, the busiest hour occured between 11:00 a.m. and 12:00 noon and corresponded to 11 percent of the daily traffic.

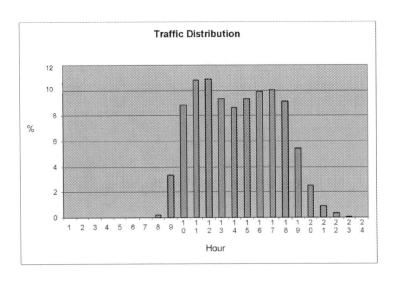

The study encompassed two different aspects: one involving negotiation of contracts and discounts and the other redesign of the network structure, including the deployment of a private network.

Negotiations

This aspect consisted of comparing the company's current costs with the market alternatives available. The current telephone bill (1-800) was calculated, verifying how much it would cost if it was charged using eight different charging plans from four service providers. This calculation demanded the use of interconnection-cost calculators—to do them manually or even using an Excel spreadsheet was virtually impossible.

These simulations showed the correlation between the amount currently paid and the savings achievable through negotiation.

Charging plan	Total value of a reference bill	Difference from the current costs	%	Average price per minute
Telco 1 (General public tariffs)	USD 2,234,624.00	USD 1,406,252.04	169.76%	USD 0.49
Telco 2 (General public tariffs)	USD 2,078,482.77	USD 1,250,110.81	150.91%	USD 0.46
Telco 3 (General public tariffs)	USD 2,054,728.00	USD 1,226,356.04	148.04%	USD 0.45
Telco 3 (General enterprise plan)	USD 1,064,075.22	USD 235,703.26	28.45%	USD 0.24
Telco 3 (specific currently used)	**USD 828,371.96**	**USD 0.00**	0.00%	USD 0.18
Telco 2 (Specific proposal)	USD 673,642.35	(USD 154,729.61)	-18.68%	USD 0.15
Telco 4 (Specific proposal 1)	USD 752,195.65	(USD 76,176.31)	-9.20%	USD 0.17
Telco 4 (Specific proposal 2)	USD 687,619.40	(USD 140,752.56)	-16.99%	USD 0.15

As can be seen in the spreadsheet above, there was some room to obtain savings by negotiating with the service providers. There was the possibility of achieving at least a 15 percent discount from the current costs.

Redesigning the structure

Although there was room to obtain discounts through negotiations, there was also the possibility of archiving savings by redesigning the network. With this objective, the traffic was mapped, analyzing the telephone bills and identifying the origination and destination of traffic (traffic matrix), verifying several alternatives including the alternative of building a private network.

The optimized structure identified included twenty-two regional nodes (as shown below) connected to Rio de Janeiro through dedicated circuits. Local numbers were adopted in each one of these twenty-two nodes. The 1-800 number would work only outside these twenty-two areas. This structure would cost USD 557,838 per month, generating a savings of 48 percent (USD 401,389) from the current expenditure.

The study also took into consideration the cost of building these nodes (considering the total cost of ownership), including the hardware lease cost and the real-estate rental costs. The twenty-two nodes were:

	Node Name	Area code	Total number of users associated with the node	Number of users local	Number of users between 50 and 100Km	Number of users between 101 and 300Km	Number of users between 301 and 700Km	Number of users above 700Km
1	BELO HORIZONTE	312	109,203	47,843		15,821	32,126	
2	JUIZ DE FORA	322	38,946	17,501		10,783	10,662	
3	UBERLANDIA	342	59,831	15,509		9,408	29,371	
4	MACEIO	822	52,363	20,072		2,699	29,266	
5	MANAUS	922	28,854	21,888		182	312	
6	FEIRA DE SANTAN	752	51,236	26,508	3,708	2,822	4,363	
7	ITABUNA	732	33,819	18,516		646	9,574	
8	SALVADOR	712	35,485	34,379		1,106		
9	FORTALEZA	852	86,730	66,795		1,174	7,682	
10	BRASILIA	612	108,283	69,901	18,785	2,849	2,666	
11	VITORIA	272	78,184	38,052		3,990	33,158	
12	GOIANIA	622	82,807	62,717		5,889	12,435	
13	SAO LUIS	982	30,017	14,532		2,339	4,147	
14	CUIABA	653	46,450	23,131		92	3,032	
15	BELEM	912	22,597	18,396	288	334	146	0
16	JOAO PESSOA	832	36,580	24,787		4,658	7,063	0
17	RECIFE	812	102,397	90,704	3,236		8,457	
18	CURITIBA	412	96,423	38,524		7,783	23,711	0
19	RIO DE JANEIRO	21	113,873	85,752	7,465	1,172	19,484	138,635
20	NATAL	842	26,537	25,189			1,348	
21	PORTO ALEGRE	512	52,124	27,666	225	3,571	15,714	0
22	SAO PAULO	11	106,302	60,125		23,184	22,993	
	TOTAL		1,399,041	848,487	33,707	100,502	277,710	138,635
	Percentage		100.00%	60.65%	2.41%	7.18%	19.85%	9.91%

The topology calculated was a star configuration, although some marginal gains could be obtained through composing flows over the same physical paths (such as traffic from Porto Alegre and Curitiba coming to Rio de Janeiro through São Paulo).

Through the deployment of tools, it also becomes possible to execute simulations identifying how much the structure would cost if the IVRs were distributed and if the local callers paid for the calls.

Service	Monthly cost
Backbone Cost	USD 116,904.71
Access cost	USD 424,983.59
Hardware cost	USD 15,950.00
Total	USD 557,838.30

As can be seen, this kind of analysis not only identified the possibility of substantial economies but also made possible the evaluation of several operational scenarios.

The savings achievable through traditional negotiation (15 percent) were way below what could be achieved by rearranging the structure. An interesting aspect of this case in particular was the fact that the organization didn't actually implement the cheapest capitation structure. What was done was that the current provider was contacted and presented with the study. The organization explained that if a substantial discount wasn't provided they would implement the project. Confronted with the alternative of losing everything, the telco gave up and provided a discount that brought the monthly value close to USD 560,000 (the organization threatened not only to build a capitation network but also to let the access cost to be paid by the user—a situation in which Telco 3 would lose almost 60 percent of the access fees).

This case gives a classical example of how a well-conducted planning project can help to extract discounts from the telco. As we saw, using a traditional approach—"kick the telco"—wouldn't have gotten much beyond 15 percent.

6.3   A major Brazilian organization replans its call-center operations

In this section, we will discuss a real case in which the modeling techniques were deployed. This case involved a large Brazilian financial-services firm—leader in traditional, personal-banking services that operated its own call centers.

The company employed about 80,000 people in 7,972 points of presence. It covered all of Brazil, providing services to more than 13,400,000 clients. The company had 32 telecommunications main nodes and spent USD 16,000,000 monthly on telecommunications.

The company was studying the possible alternatives to centralize its call-center services that were distributed among ten different sites. The idea was not to physically unify the structures but rather to integrate them and make them cheaper, more accountable, and more manageable. The organization was also evaluating the feasibility of introducing new services.

The organization had difficulty knowing what would be the ideal structure to support the traffic and to establish a model in which all the costs involved in implementing a new service were easily identified.

The impossibility of defining the correlation between the costs to implement a new service and the revenue that would be generated by it was paralyzing the internal decision process. Consequently, a model was needed that could establish the correlation between service cost, service revenue, and percentage of clients using the service.

The interconnection costs available were from the Brazilian main telcos for data and voice prices. As a reminder, in Brazil, calls inside the same area code (local calls) are charged and 1-800 calls can have different charges, depending on where the call is originated.

The model contemplated two situations: the client paying for the access and the organization paying for it.

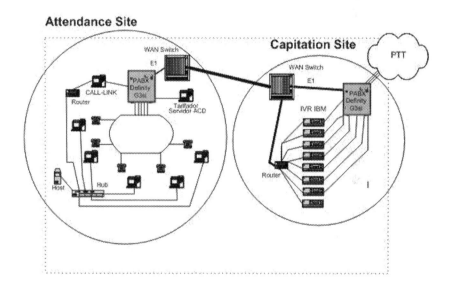

The equipment used as reference to build typical nodes were PBXs, WAN Switches, routers, and IVRs.

Grouping the users by area code and varying the numbers of nodes, we managed to identify how the percentage of the users varied when we varied the number of aggregation nodes.

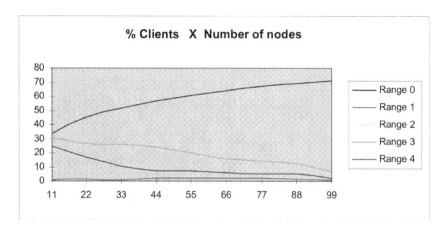

The ideal capitation structure was identified with a topology using forty-eight traffic capitation nodes.

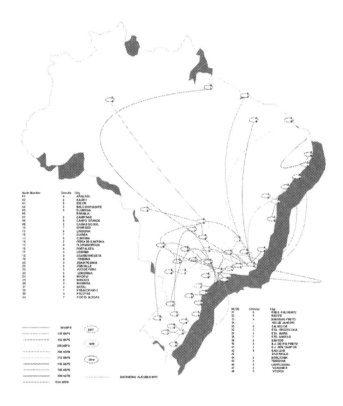

The geographical dispersion and the modeling of volume per service made it possible not only to identify the ideal capitation structure but also to create a model in which we could vary the volume and check how much the structure would cost for each volume scenario.

Armed with these analyses, the organization was able to optimize its call-center structure and, most importantly, to decide which service was worth implementing. It became possible to evaluate the impact of each service on the total cost of the structure.

In addition, the model was adapted to include the revenue generated by each service transaction, allowing the organization to establish a correlation between volumes/costs and volumes/revenues, consequently identifying the minimum percentage of clients necessary to make each service feasible.

This project changed the initially proposed model to integrate the actual call-center services, ensuring a cost reduction from USD 6,000,000 to USD 500,000 monthly by developing a distributed network with different topology adapted to Brazilian tariffs.

Maybe the biggest achievement of this project wasn't so much identifying a more cost-effective solution as it was providing the corporation with an instrument that would allow them to decide which services to implement.

The importance of having this information is often overlooked by organizations. Typically, the problem is considered too complex to be modeled, and consequently, a complete study isn't even attempted. That paradigm was challenged in this project.

Chapter 7: GSM Gateways

When dealing with traffic destined for mobile phones, it is always good to have what is known as "on-net" calls. That means calls that go all the way from the caller to the receiver within the same provider network. This guarantees quality and low costs. In some countries, the cost to make a call from a fix line to a mobile one can be very high, and even without originating the call from a trunk belonging to the same provider as the receiver (on-net), just originating the call from a mobile trunk (from another mobile provider) is enough to reduce substantively the cost of the minute.

Therefore, in order to reduce the costs of outgoing calls to mobile phones, a very common strategy is to redirect the calls to dedicated mobile trunks. To be able to do this redirecting, you have to have a least-cost routing process in place. The strategy becomes even more cost effective if you manage to originate the calls from a mobile trunk belonging to the same operator to which the call is designated (on-net calls).

Typically you have two main strategies: having dedicated trunks linking your switch directly with the mobile-phone operators or having devices called "GSM gateways," which are machines with mobile-phone chips installed and that work as if the calls were generated from mobile trunks—as in fact they are.

This kind of possibility—reducing costs by redirecting calls through a mobile gateway and selecting the trunk originating the call based on the destination of the call, therefore making it an on-net call—adds to the problems of doing the least-cost routing directly by the dialers. This is because in large installations with a wide variety of types of mailings it is very difficult to equip each dialer with the adequate quantity of mobile trunks to support the necessary volume of calls to mobile phones in all moments. It calls for a trunk group for each mobile operator, plus the fix trunk groups in each dialer.

This is the reason why it makes sense in large installations to have machines (usually tandem PBXs) pooling all trunks to be used by all dialers and doing the call routing. In this scenario, the possibility of having all trunk groups of specific type and provider busy at the same time is very low.

A typical arrangement would either implement the least-cost routing through a PBX or use the PBX only to separate preliminarily the calls designated to mobiles and forward them to the GSM gateway and then the gateway itself would do the selecting of the right trunks group to make the calls on-net.

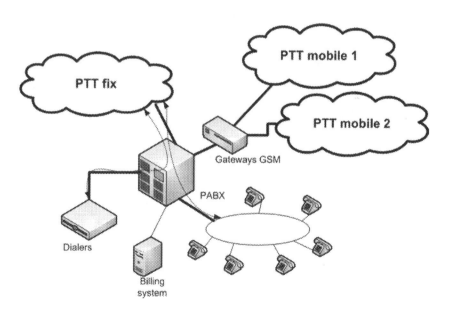

In addition, many providers offer the possibility of getting special prices and even completely free calls between the mobile trunks belonging to the same organization—what we could call "on-group."

That opens the possibility of not only reducing the cost of talking with mobiles, but also talking for free (or for extremely low fees) with mobiles belonging to the same organization, including sites equipped with mobile trunks (directly or through GSM gateways). That is a particularly attractive possibility, considering that in most organizations a large percentage of the fix-mobile traffic occurs between the fix lines and mobiles belonging to the organization itself.

On top of that, very often you may attain big advantages associated with using mobile trunks between areas where the fix lines demand change in area codes and the mobile network doesn't.

Here, we are going to show how a study could be conducted to verify the feasibility of using GSM gateways. The first step is to separate the calls made to mobile phones by mobile operator and by type (VC1-Local, VC2—Intra-state, and VC3 inter-state):

Mobile provider	Type	Quant	Minutes	Current cost	% quant	% Minutes	% Cost	Cost of the minute mobile to mobile		Cost if using GSM gateway	
Telco 1	VC1	7,680	5,177.53	USD 2,780.77	1.23%	1.11%	0.69%	USD	0.32	USD	1,656.81
Telco 1	VC2	263	223.20	USD 203.14	0.04%	0.05%	0.05%	USD	0.55	USD	122.76
Telco 1	VC3	2,175	2,217.04	USD 2,063.19	0.35%	0.48%	0.51%	USD	0.55	USD	1,219.37
Telco 2	VC3	2,132	1,556.31	USD 1,491.75	0.34%	0.33%	0.37%	USD	0.55	USD	855.97
Telco 3	VC3	13,804	8,171.50	USD 7,826.88	2.21%	1.75%	1.94%	USD	0.55	USD	4,494.32
Telco 4	VC1	32,132	22,697.57	USD 12,346.44	5.14%	4.87%	3.06%	USD	0.25	USD	5,674.39
Telco 4	VC2	10,561	8,692.07	USD 8,312.07	1.69%	1.86%	2.06%	USD	0.25	USD	2,173.02
Telco 4	VC3	75,390	60,411.59	USD 57,841.04	12.07%	12.96%	14.36%	USD	0.25	USD	15,102.90
Telco 5	VC3	7,664	3,758.75	USD 3,595.70	1.23%	0.81%	0.89%	USD	0.55	USD	2,067.31
Telco 6	VC1	11,417	8,521.23	USD 4,634.48	1.83%	1.83%	1.15%	USD	0.28	USD	2,385.94
Telco 6	VC2	294	306.57	USD 293.62	0.05%	0.07%	0.07%	USD	0.28	USD	85.84
Telco 6	VC3	53,875	44,907.06	USD 42,986.81	8.62%	9.63%	10.67%	USD	0.28	USD	12,573.98
Telco 7	VC3	294	156.73	USD 149.97	0.05%	0.03%	0.04%	USD	0.55	USD	86.20
Telco 8	VC3	8,316	7,321.87	USD 7,000.83	1.33%	1.57%	1.74%	USD	0.55	USD	4,027.03
Telco 9	VC1	24,067	21,296.84	USD 11,576.84	3.85%	4.57%	2.87%	USD	0.19	USD	4,046.40
Telco 9	VC2	3,337	2,870.22	USD 2,744.91	0.53%	0.62%	0.68%	USD	0.21	USD	602.75
Telco 9	VC3	49,332	46,146.30	USD 44,139.32	7.90%	9.90%	10.96%	USD	0.21	USD	9,690.72
Telco 10	VC1	39	21.83	USD 11.76	0.01%	0.00%	0.00%	USD	0.55	USD	12.01
Telco 10	VC2	8	1.64	USD 1.59	0.00%	0.00%	0.00%	USD	0.55	USD	0.90
Telco 11	VC1	75,205	46,694.44	USD 25,407.22	12.04%	10.02%	6.31%	USD	0.18	USD	8,405.00
Telco 11	VC2	20,145	13,290.60	USD 12,711.32	3.22%	2.85%	3.16%	USD	0.36	USD	4,784.62
Telco 11	VC3	226,578	161,774.21	USD 154,709.81	36.27%	34.70%	38.41%	USD	0.36	USD	58,236.72
Total		**624,728**	**466,215.10**	**USD 402,829.44**	**100.00%**	**100.00%**	**100.00%**			**USD 138,306.96**	

Once we have a clear view of the volume per type terminated in each mobile provider, we can calculate the cost of the calls if they were charged using the "on-net" tariff of each provider. Comparing this total with the current costs, the savings attainable through using mobile trunks and routing the calls properly becomes clear.

As we can see in this example, if we configure the equipment to route the calls designated to the service providers telcos 4, 6, 9, and 11 to gateways GSM, the monthly savings would be around USD 264,000 (a reduction of 65 percent of the current costs).

Of course, we have to discount the cost of the equipment. In this case, the GSM gateways would cost USD 4,000 monthly, which would leave us with a net saving of USD 260,000 a month. In some cases, we may have some additional costs such as maintenance of the routing database and mobile network access.

The large savings achieved in this particular example is not unusual. In addition, we have to remember that in this example we are not discounting the calls, which are designated to mobile phones belonging to the organization itself. If we separate these on-group calls and get a special discount for them, the savings could be even bigger.

This kind of strategy is not always applicable. In some countries, it is even forbidden by the regulators. Another common problem is the limitations of the local ERBs of some of the mobile providers, which limit the deployment of the GSM gateways. For example, the provider's ERB near the datacenter may not support a thousand mobile trunks working 100 percent of the time. Nevertheless, the gains attainable through this kind of strategy are so substantial that it is worth investigating all these details.

Chapter 8: Billing Systems

The billing system is a very important component in controlling a call-center operation. Through the billing system, it becomes possible to control the operation of the infrastructure by measuring the traffic volumes and costs. Therefore, putting in place an effective and precise billing system is crucial. It is the tool that will give you the basic information about your traffic, and it is through it that you will be able to follow your telco costs by type, spot discrepancies between what was actually charged and what was supposed to be charged, and of course define the telco cost of each operation that, depending on the contract, may or may not result in reimbursements from your customers.

Basically, the main objective of a billing system is to guarantee an adequate level of control over the telco costs. To achieve that, though, some considerations are necessary:

- You must guarantee that all trunks are connected to devices from which logs are collected and processed. If this isn't the case, you will have calls on your bill that are not verifiable by your billing system.
- All the process of collecting the logs (CDRs) must be automatic and absolutely without human interference; otherwise, the process may not be reliable.
- The team in charge of the billing system must have restricted access to the telco contracts and the tariffs and keep the database of the billing system strictly updated.
- The team in charge of the billing system should be the same as that in charge of defining the least-cost routes and verifying the bills. This guarantees that the information related to tariffs, least-cost

routing, and bill totals per provider are thoroughly followed and the relationships of cause and effect are clearly perceived. To be successful managing the traffic and guaranteeing lower operational cost, it is crucial to have the same group of professionals seeing and understanding the costs (billing) and effectively acting on them (least-cost routing).

Usually in large call-center service providers, the billing system is also the tool used to separate the traffic by specific operations in order to allow this traffic to be charged back to the clients (not necessary charging the client exactly the same amount charged by the telcos).

A complete billing system usually encompasses devices to collect and store the call logs and servers to process and present these data. In large and heterogeneous environments, implementing effective billing systems isn't a trivial task.

First, you have to make the call-log layouts standardized, because different equipment may generate logs with different formats—all of them have to be normalized in only one layout used by the billing software.

A second issue, very often overlooked, is processing logs of equipment interconnected without duplicating calls. A clear understanding of the problem can be achieved in the following example:

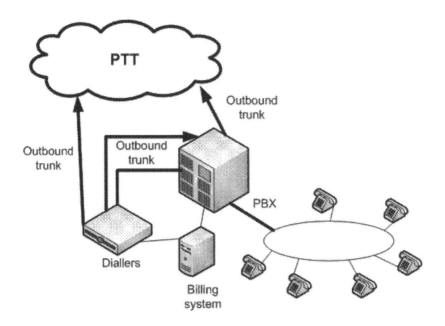

As you can see, you may have calls originated from the dialers that go directly to the public network and generate logs only in the dialer, and you may have calls originated by the dialer going to the public network through the PBX, which generates logs in the PBX and in the dialer. You should be able to separate these calls and avoid having the same call be identified as two by the billing system. This situation is even trickier when we can't just ignore the logs of one of the devices. You may need information about the extension from the PBX device and the information about the mailing from the dialer device, or you may have calls generated directly by the PBX.

In addition to the difficulties associated with the initial installation, keeping the billing system working correctly demands a considerable effort to keep the databases updated with the right tariffs, trunks per equipment, and list of extensions, to mention just some of the complications.

Another important aspect is the need for good synchronism between the installation teams and the team in charge of the billing system. It is very important to have a process in place that guarantees that any new equipment installed be linked to the billing system, and that each time a trunk is installed/de-installed or shifted in any device this information must be forwarded to the billing system team which will update the system database.

If there is no process in place to guarantee the connection of new devices to the billing system or detailed trunk identification, you may have a situation in which some calls are not collected or if collected, identified and priced as going through a specific telco when in reality they went through another. Only by being careful about this process can you guarantee that your data reflects the reality.

It is also very important to separate the trunk groups in the billing system exactly as they are separated in the bills in order to allow you to generate a mirror of the bill and compare it with what you actually were charged.

And you need a mechanism to keep the tariffs updated, and in case a tariff changes, it is important to have a clear identification of the date from which the new tariff applies.

Ideally, the billing system should be centralized and be able to collect and process the billing information of all sites and all devices.

The billing system must be able to identify the following incoming and outgoing call parameters (including the unidirectional and bidirectional trunks) and generate reports showing:

- Number originating or receiving the call (local, national, and international)
- Number of the extension that originated or received the call
- Date of the call (day/month/year)
- Time the call started
- Call duration
- User ID and original extension
- Register of transference

The billing system must be able to automatically and periodically (with the periodicity defined by the user) generate at least the following reports:

- All outbound calls (internal, local, national, and international)
- All inbound calls (through direct access trunks)
- Calls by extension, indicating the external number dialed
- Calls whose duration exceeds a given time (inbound and outbound)
- Calls executed through the operator consoles

Ideally, the system should be able to control the calls originated using user IDs, regardless of the extension actually used. In addition, the system should be able to control calls transferred, generating a specific log for each transference.

In terms of guarantee of operational continuity, the billing system (here including the whole apparatus) must be able to insure the continuity of the process of billing regardless of minor failures. Therefore, the structure should be designed to store the call logs for some reasonable period of time if the connection with the billing workstation becomes unavailable.

The designer of the structure must guarantee that all devices being billed are equipped with internal storage capacity to support at least eight hours (typical) of operation without the need for unloading the logs to the billing-storage system.

For the most important devices, you may consider using external buffers able support the logs for additional hours.

The system must have mechanisms to identify clearly if some device failed to provide logs during some period of time.

When defining the requirements of the system, you should guarantee that it complies with the following:

- Adaptable to different configurations and needs
- Have an acceptable performance during the highest traffic hour
- Be flexible to comply with changes and upgrades

You may also specify the operational system and database to be used by the billing software itself.

In large installations it is advisable to divide the billing system functions into two independent modules: data collection/storage and management system. It is a good practice not only because by doing so you can separate the functions into two devices and may use the duplication to implement backups (one device may be able to execute the other function in case of need), but also because through this strategy you may be able to operate the management system without loss of performance during the log collection and processing.

The data collection process must be able to receive and store all logs generated by all network modules. The management system in its turn must be able to treat the data collected, allowing emission of reports, inclusion/exclusion, and editing of data and configuration of parameters, such as tariffs.

You should make sure that the billing system is able to collect the logs from the devices using different strategies: direct connection, connection through a LAN, or dial-in. The data collection must happen automatically and independent of the model of the equipment in each site. The provider of the billing system must be able to adjust all the necessary interfaces and make all the necessary file layout adjustments.

If you have sites located in many different countries, the billing system must be able to treat more than one currency, converting and consolidating costs.

When procuring a billing system, you should provide the vendors a general view of the volumes involved and the number of calls generated by month and by day. Based on this information, the vendor will specify the hardware capacity required and the maximum number of calls that the system can handle (some spare capacity is advisable). Therefore, the vendor must state the limits of the proposed solution and the alternatives to expand it. Remember, it is your responsibility to tell them your volume; they don't know that.

You should emphasize to the vendor that the proposal should include all devices and software associated to transport the logs from the devices to the billing workstations. That means if you define, for instance, that the logs will be collected through the company LAN, if a new router is required it should be included in the proposal (all additional hardware and software necessary to the implementation/operation of the billing system as a whole).

A very useful feature, often overlooked by the telecom managers when preparing technical specifications of billing systems, is the remote access. Ideally, the billing system must be remotely configurable (including all parameters). This feature must include all sites and the main unit. The remote access must allow at least the following features:

- Set of date and hour
- Switch specific parameters
- Set of data transference time
- Access through dial-in
- Verification of the level of storage and memory occupation set of the data-collection mode (on demand or automatic)
- Adjustments in the data format conversion (text to the database used by the billing system)
- Access to some billing reports

When defining the requirements of a billing system, it is important to make sure that it can be accessed through a network, and you should be able to expand it, using as many workstations as the increase of the volume may demand—it is good to know in advance the expansion cost. In other words, the system must be scalable in all its features, and you should know how much it will cost to expand.

Any system must have access control through name and password, and the users may be divided into different levels (preferentially with possibility of defining the user accessibility individually). In addition, the system should be user friendly, which makes the training easer, and all reports should be seen in the screen before being printed using standard WYSIWYG (What You See Is What You Get).

The ideal system should have a report generator with filters to select the calls, which match a particular set of requirements (data, cost, duration, extension). In addition, it would be ideal to be able to save the report formats in some sort of report gallery.

The report generator should be able to provide reports of the calls by at least the following criteria (detailed or grouped):

- Costs centers
- Extension
- Trunks and tie-line
- Special tariffs
- Number dialed
- External telephone list
- Type of expense codes
- Additional fees
- Tariffs
- Users
- Operators desk
- Tariff per service provider
- Per duration or cost above a given value
- Forwarding prescheduled reports automatically through defined e-mails.

Another interesting feature to have is the possibility of storing a telephone number and then generating a report indicating how many times this particular number was called.

The system must be able to make security backups automatically through the network and allow open database exportation for different formats such XLS, PDF, DBF, or TXT, allowing easy integration with external systems.

Another important function is the simulation and recalculation of logs based on different set of tariffs.

Ideally, the system should allow audit trails that indicate all interventions and processes operated by the system. That means indicating things such as what each user did to what treatment each billing log (CDR—Call Detail Report) received.

8.1   Effectiveness of the billing system

The effectiveness of a billing system very often has more to do with the way it is installed and operated than with the software itself. That said, let's analyze the main aspects that determine effective control of the bills:

Aspects linked with the installation:

1) Depending on how the hardware is connected, you may have problems collecting all logs automatically and have to add part of the logs manually (even if only eventually). In addition, if you don't have thorough control, should any device fail to send its logs, you may end up generating reports that don't reflect what really happened.

2) In some situations, when a device is interconnected with another, you may have situations where the same call generates logs in more than one device. This has to be very carefully controlled. It is worth mentioning that in some circumstances you cannot just ignore the logs of one of the devices, because these logs may have information that is relevant (for example, one device generates the call and knows the extension used, another forwarded the call to the public network and knows which provider was used).

3) Other common problems are associated with wrongly documenting the trunks and therefore wrongly identifying the calls. If the identification of the trunks isn't kept updated, the billing system will identify the calls that originated through a specific trunk as being supported by one service provider, when in reality it was supported by another. That means the cost of the call is wrongly calculated. This is a common problem and occurs due to a failure to communicate between the installation teams and the billing team.

4) Very often, the trunks are not grouped exactly as the service providers charge them. This prevents the organization from generating a truly mirrored bill, which in turn jeopardizes the cost control.

5) Very often, the team in charge of the billing process is not linked or associated with the professionals in charge of the negotiations and contract management. This generates situations in which

new tariffs or billing rules are negotiated but the billing team is not informed.

6) In the same way, there are situations in which the billing team isn't linked with the call-routing team, and this renders the management of the call routing ineffective since the team able to route the calls properly doesn't know how it should be done. Sounds strange, but it is a very common scenario.

Chapter 9: Billing Auditing

Billing auditing is the process of verifying carefully in the bill if what was charged by the service provider was what was agreed to in the contract. This verification includes the recalculation of each call and the verification of every item charged in the bill.

Typically, most organizations don't do the complete recalculation of all calls every month; usually they do that only for the ones in which big discrepancies between what was identified by the billing system and what was actually charged were spotted.

Nevertheless, verifying the bills is a necessity in a large call center. Verification identifies charge discrepancies and enables the organization to get the due reimbursement for overcharging from the PTT providers. Experience shows that savings between 5 percent and 12 percent are usually attainable through regular auditing of the bills. Considering the expenditures with telecom of large call centers, 5 percent can mean an enormous amount of money, and therefore auditing the bills usually is worth the effort.

Here it is interesting to mention that when we say verify the bills most people think about re-calculating the value charged for the calls. Although this is no doubt an important part of the verification, there are several other things that should be checked:

- Verification whether the charges are associated to trunks that actually belong to the organization
- Verification if there are undue charges for installation or trunk subscription

- Verification if there are penalties for late payment unduly charged
- Verification if there are charges for not achieving the minimum-committed volume

Whether the charges are associated to trunks that actually belong to the organization should be verified. Although it may seem a bit obvious, the first and basic verification (very often not executed) is to check if the bill belongs to the organization, and the resource is actually being used. It is not uncommon to receive bills that don't belong to the organization or previously belonged but the resource was cancelled or deactivated. In large operations with several sites (sometimes in different countries), just to be able to check if the bills belong to the organization and verify if the amount charged is compatible with the historical value is in itself a difficult task.

Verification if there are undue charges for installation or trunk subscription is a very common problem. That includes all sorts of charges ranging from installations fees to trunks subscriptions and special services. Those charges sometimes are a large percentage of the bill and their verification is far from easy.

Verification is necessary if there are penalties or interest for late payment unduly charged. Such charges very often are wrongly calculated and don't correspond to what was defined in the contract. The bill verifier has to check if the penalties and interest are due and if they were rightly calculated as defined in the contract.

Verification if there are charges for not achieving the minimum-committed volume is important to verify if there are charges for not achieving the minimum-committed volume and if these charges were properly calculated as defined in the contract. Usually, the difference between what was committed and what was actually used has to be paid in full.

These kinds of problems very often provoke more overcharges than the mistakes in the calculation of the calls.

Besides the direct financial gains, auditing the bills enables the evaluation of many other important aspects linked with the management of a large call-center operation:

- Identifying the delicate balancing act between the minimum volume commitment and tariffs
- Identifying volume and duration of the calls, which is crucial to negotiate charging granularity
- Identifying the point at which least-cost route configuration in the PBXs needs adjustment due to changes in tariffs
- Knowing when it becomes feasible to have a private-voice network (interest of traffic concentrated in some specific area code)
- Accurately calculating the number of trunks and circuits (capacity planning)

Analyzing the voice bills enables accurate answers to all these questions, and the telecom manager should be aware that the analysis of a telecom bill is much more than just checking the tariffs. It is also about verifying the traffic and comparing the current prices against available alternatives.

The process of verifying the bill usually follows these steps:

- Verify if the bill and the resources belong to the organization.
- Verify if the value charged is compatible with the historical value.
- Verify if the minimum-committed volume was achieved (if not, recalculate the penalty and check if it was rightly charged).
- Verify if there is any penalty and interest for late payment being charged.
- Verify if the charging period is correct and if there is any additional fee being charged for unsolicited services, such as installation or subscription.
- Identify resources not in use, such as trunks without any calls.

After verifying these basic items, you should check the calls, recalculating the value and checking the traffic interest:

- Verify from and to where each call was made.

- Identify the area codes from and to each call.
- Verify how much each call is supposed to cost, based on the organization's specific contract, and identify the discrepancies.
- Identify if the taxes were properly applied.
- Verify if there are calls charged outside of the admissible charging period.

When auditing telephone bills, it is important to consider that this procedure is effective only when there is a process in place to define how and when the organization will be reimbursed. Contracts must have dispute clauses, and, ideally, the contracts in place should foresee that. If errors were spotted in the bills before the payment due date, the organization can notify the service provider and pay only what was considered due. The values over which there is disagreement must be discussed jointly. If the charges prove to be right, the organization pays the service provider without penalties for delaying the payment (interest is due).

The organization musts define a formal process through which all invoice disputes are treated. The process needs a definition of time frames for each party involved and should mirror the dispute resolution clauses of individual contracts or master agreements. Typically, the whole process of disputing a bill from identification to solution should not exceed three months.

The SLA must consider that the invoice payment doesn't imply acceptance of the charged values. Ideally, the organization should have at least one year to audit the values charged, and the SLA should foresee that. If the errors identified exceed 5 percent of the total value of the bills, the service providers must reimburse the organization for the costs involved in auditing the bills. This is a considerable cost and a good mechanism for keeping service providers accurate.

The organization must define clearly the time after which charges are not acceptable (in some countries it is defined by law). For example, services provided more than six months ago should not be charged.

The organization must schedule regular meetings with the service provider invoice team. Such meetings should be forums to discuss problems with the invoices and items added, changed, and cancelled.

The organization should try to define standardized invoice cycles and guarantee that all invoices are due on the same day; this simplifies the billing and payment process.

Of course, all these recommendations depend on negotiation at the contract stage and arguably these points belong in the negotiations chapter. However, these clauses only become meaningful if actual billing verification is done; without it you will have no recourse or knowledge of billing errors.

Chapter 10: Dialers

The outbound operations usually are extremely challenging for telecom managers. These are the kind of operations in which just keeping the lights on is not enough, and usually the difference between a good installation and a bad one can make a huge difference in terms of productivity and influence directly the company's bottom line.

The deployment of devices called diallers (British English) or dialers (American English) designed to automatically dial telephone numbers, can generate significative productivity gains. However, you should not have the delusion that just buying the devices generates the productivity gains by default. The telecom manager has to understand that the productivity gains come from a well-balanced combination of three factors:

- Well-chosen equipment
- A good installation and proper configuration
- A good management of the tool in the day-by-day operations

Dialers are crucial tools for outbound call centers, because they free the agents of the work of dialing and make it possible for them to concentrate their effort in actually talking with the clients. If badly implemented, dialers may produce lackluster results and increase operational costs.

In addition to the basic function of dialing and transferring the call for an human attendant, the dialer device can also perform several other tasks, such as announce verbal messages (called a robocall in the United States) or transmit digital data (like SMS messages).

In addition, these devices usually can verify the dialed numbers and change them to seamlessly provide services that otherwise require lengthy access codes to be dialed.

Typically, a dialer can automatically insert and modify the numbers depending on the time of day, country, or area code dialed, allowing also the selection of the service providers who offer the best rates. For example, a dialer could be programmed to use one service provider for international calls and another for mobile calls. These processes are known as prefix insertion and least-cost routing.

Although most dialers can execute the least-cost routing, it is interesting to note that in large installations it may not be ideal to do so because that usually means the number of trunks (T1/E1) supported by each dialer box may not be enough for all dialing circumstances. For example, if in a particular moment you have lots of fix-to-mobile calls, the number of mobile trunks available in each box may not be enough. This is the reason why it usually makes more sense to pool the trunks groups in a large voice switch and do the routing from there.

Dialer systems are commonly used by telemarketing organizations involved in B2C (business-to-consumer) calling, because it allows their sales representatives to have much more customer contact time. Market-survey companies and debt-collection services that need to contact and personally speak to a lot of people by telephone may also use dialers.

More commonly, dialers are now being used as a quick and easy way to automate all types of calls that would otherwise be made manually by a call center, such as welcome calls for new customers, customer service call backs, appointment confirmations/reminders, or even for the automation of large numbers of ad hoc calls that might need to take place (such as by a taxi company or a parcel-delivery service).

The basic idea behind the device is that if a person were to sit down and manually dial one thousand people, a large percentage of these calls will not result in contact with someone at the other end. Out of one thousand calls made, typically only about 25 to 35 percent would actually connect to a live person. Of the rest, a large number (often 40 to 60 percent) won't be answered at all, around 10 percent might be answering machines, faxes, modems, or other electronic devices, around 5 percent of numbers would be busy, and the rest will result in network errors or be identified as invalid numbers.

For call centers that need to make large numbers of outbound calls, this represents a big problem. Typically, in manual-dialing environments, any given agent will spend around 90 percent of his or her time listening to the phone ring, waiting to talk to someone, or dealing with invalid numbers, or answering machines, and only about 10 percent of the time actually doing what they are really there to do.

The benefit of dialers in general is that they can make many more calls in a much shorter period of time than an agent manually dialing each phone number. If the dialer encounters a busy signal or no answer, it will schedule itself to dial the number again later without human intervention. The system can also keep track of an entire campaign's progress in real time—which would be nearly impossible if attempted manually.

10.1 Predictive dialers

An evolved form of automated dialer are called predictive dialers. In addition to automatically dialing a list of phone numbers for customers or prospective customers, as traditional dialers do, the predictive dialers execute two sets of activities, which increase enormously the productivity of the dialing process:

1) Screens out no-answers, busy signals, standard information tones (SITs), and fax/answering machines, only sending calls that reach a live person to a sales agent. Here it is interesting to note that to be able to screen such things it is necessary to have a fine tuning between the public network signalling and the device. That means lengthy installation times and a very high level of sensitivity in any shift of trunks and operators.

2) Using intelligent algorithms, these devices can detect when an agent is wrapping up a call; they'll then begin dialing the next

number and send the call to that agent as soon as a dialer reaches a live voice on the other end. These algorithms are also capable of detecting the number of available telephone lines, available operators, and average length of each call. Therefore, these devices predict the dialing rate in order to match it with the number of available agents. While the basic autodialer merely automatically dials telephone numbers for call-center agents who are idle or waiting for a call, the predictive dialer uses a variety of algorithms to predict both the availability of agents and called-party answers, adjusting the calling rate to the number of agents it predicts will be available when the calls it places are expected to be answered.

By using a dialer to filter out these unproductive calls and to spare the agent from having to wait (adjusting the dial rate), call centers can improve productivity enormously. Agents can now spend on average around 80 percent of their time talking to customers and only about 20 percent of their time waiting for the next call.

Compared with the conventional dialer (known in some countries as autodialer or power-dialer), predictive dialers have shown increases in talk time from twenty minutes in the hour to almost fifty (25 percent to 85 percent). However, predictive dialers are more suitable for low-quality mailings and large numbers of agents, and we should be aware that an unexpectedly high contact rate can overwhelm the system, leading to a high rate of call abandonment.

We should be aware that the ability to identify bad numbers, answering machines, and fax machines isn't what defines a predictive dialer—sophisticated power-dialers may do those things also—but rather it is the ability to use this information, plus the average duration and available agents, to adjust the dial rate.

In operational terms, we have to understand that before running a campaign, the call list data (this is usually called a mailing or campaign), is loaded into the dialer. Then, the dialing process starts, and statistics are kept about what is happening. Most predictive dialers generate reports that indicate call attempts and unsuccessful calls by type. Unsuccessful calls are often analyzed to determine if the number called needs to be called back later or needs special treatment, such as a manual or autodialed call by an agent to listen to an answering-machine message.

Some companies adopt the practice of what is called "cleaning the mailing," which consists of using dedicated dialers to call the numbers very quickly. The call duration is just enough to check if the number is valid (whatever the reason) and separate it from the mailing. Only after performing this cleaning are the separated numbers loaded in production dialers. Usually, the devices used for cleaning the mailing lists are connected to service providers whose charging granularity doesn't exceed 18s as the minimum time charged per call, regardless of the real duration of the call. This practice maximizes the productivity because it allows the separation of the different kinds of situations in a more dedicated form. As examples, fax machines may receive a predefined fax message, and answering machines may receive a prerecorded message. Of course, it could be done by the same dialer in the first place, but very often infrastructures are not uniformly built and, depending on the situation, this strategy may be very useful.

10.2   Autodialers versus predictive dialers

Power-dialers are usually best suitable to business-to-business (B2B) applications, in which more ability from the sales agent is required to navigate receptionists, auto attendants, and voice-mail systems. No current dialer technology is able to navigate the multiple options of a voice menu in a business environment without the guidance of a live representative.

Differently from the predictive dialer, a power-dialer automates the dialing functions in a very simple way by just dialing sequentially a list of numbers. Here we may differentiate power-dialers from a solution widely known as preview: the term "preview" usually refers to a solution in which each user commands the dialing process—almost as a speed-dialing process. The power-dialer, on the other hand, generates the calls independently of the user command and has a ratio of number of user versus number of trunks other than 1:1.

Talk time and total call attempts are not quite as high as a predictive dialer; however, if you deploy a power-dialer in a B2B or multiple calls B2C environment and associate it with guided-voice messaging technology and a CRM or lead management database, the power-dialer shows its full potential.

Power-dialers are best suitable for telesales, inside-sales, or outside-sales organizations in which the sales process is more complex and requires multiple calls for closure. It is also optimal for organizations that invest heavily in their lead sources and can't afford to abandon a single call with the resulting negative perception from the person at the other end. This translates to all B2B environments and specific B2C call centers that invest in high-value leads.

Predictive dialers, on the other hand, are the best alternative when volume of calls and time on the phone are the main requirements. The predictive algorithm requires a group of agents large enough to be able to effectively leverage the "predictive" effect with multiple lines. The group of agents concurrently logged must be at least twenty to thirty. If this is the case, there are case studies where predictive dialers are able to keep agents productively talking on the phone from forty-seven to fifty-two minutes of an hour.

Predictive dialers are specifically designed for business-to-consumer applications because they require short and consistent lengths of calls and direct-dial phone numbers. They are best when used in a one-call close type of telemarketing, short surveys, or prequalification of consumers. They are not designed to navigate the phone systems, auto attendants, or receptionists of a business environment. The predictive dialers are not suitable for a B2B environment, and even in a B2C environment it is not advisable to deploy predictive dialers for expensive or valuable leads.

Predictive dialers should not be used with leads that require multiple attempts to ensure contact. They are best for single-script, telemarketing operations and are rarely effective in telesales or inside-sales organizations that require a multiple-call or complex sale.

10.3   How a predictive dialer works

In a conventional dialer system, if you have a hundred agents working on it, for example, the dialer will dial a number of calls sometimes crudely based on the phone line to agent ratio—let's say two to one. This means that for each available agent, the system will dial the phone numbers of two potential customers. The dialing ratio doesn't change.

In a predictive-dialer system, the dialer will monitor each call and determine the outcome of the call. The system will immediately strip out any unproductive results, such as busy calls (these are usually queued for automatic redial), no answers, and invalid numbers, verify the number of agents available and the typical call duration, and adjust the dial rate. Some predictive dialers incorporate also fax/answering machine detection, which tries to determine if a live person or a machine picked up the call. Today, most dialers have some sort of recognition feature; however, be aware that tuning the device to the public network is the key to making it work properly.

The adjustment of the dial rate is necessary because if not enough calls are made ahead, agents will sit idle, whereas if there are too many calls made and there are not enough agents to handle them, then the call is typically dropped. A predictive dialer system will adjust the dial rate appropriately to avoid both situations.

An advanced predictive dialer determines and uses many operating characteristics that it learns during the calling campaign and adjusts automatically to the pattern of an ongoing campaign. Examples of such statistics include call-connection rates (both current and average for recent past days by hour of the day), average agent connection time, and geographic location dialed. It uses these statistics continually to make sophisticated predictions so as to minimize agent idle time while controlling occurrences of nuisance calls (and consequently dropped calls), which are answered calls without the immediate availability of an agent. An advanced predictive dialer can readily maintain the ratio of nuisance calls to answered calls at less than a fraction of 1 percent while still dialing ahead. However, this level of performance may require a sufficiently large critical mass of agents. Conversely, it becomes increasingly difficult to maintain a high talk-time percentage with a lower number of agents without increasing dropped calls.

10.3.1 Dialing rate asynchrony

When there is an asynchrony between the dial rate and the available agents, you have a situation with two possible outcomes:

1) There are more live parties on call attempts than there are agents available.
2) There are more agents available than there are live parties on call attempts.

Usually, when you have a situation like that, both situations occur intermittently.

If we have more live parties on call attempts than there are agents available to take those calls, the dialer will disconnect or delay distribution of calls that cannot be distributed to an agent. This is known as a silent call or a nuisance call. The called party hears only silence when the predictive dialer does not at least play a recorded message.

The experience for the clients who receive a silent call can be very unsatisfactory when we have an appreciable period of silence before a call is routed to a sales representative. This annoys people and also gives them a chance to hang up. A high rang-up/dropped calls rate is a clear indicator of asynchrony between the dial rate and the availability of agents. This is a big problem and we should be aware of the following facts:

1) A very small percentage of the mailings are actual clients, and reaching them and then having no way to treat them is very disappointing from the company perspective.
2) The client gets upset for getting a silent call.
3) These calls have a cost which, depending on the hang-up rate, can be very significant.

Some countries even regulate the number of silent calls that a company can make within a certain time frame. A good reference point on this problem is that a maximum of 3 percent of the calls, measured as a percentage of live calls made, may be dropped. More than that, and you may have a problem.

In some countries there are regulations defining the need for a mandatory abandon message to be played when no agents are available and there is an obligation to inform the caller ID.

If you have more agents available than there are live parties on call attempts, you will have agents idle, which reduces your productivity.

Therefore, it is important to keep in mind that the asynchrony causes problems in both ways. If your predictive dialer is not able to adjust the dial rate properly, you will have moments in which you will have more calls than agents (silent calls and hang-ups/dropped calls) and moments with more agents than calls (low productivity and idle agents).

It is worth mentioning an aspect, which, although operational, tends to have a big impact in the dialers' productivity: if you mix databases of numbers already used several times (in which the concentration of bad numbers tends to be higher) and new databases in the same campaign, the results tend to be worse than if you keep them separated.

10.3.2 Strategies of deployment

Today, in most countries, you have the possibility of outsourcing the dialing services to specialized companies, You may have several possible arrangements, which can run from a total outsourcing of the infrastructure (PBX, ACD, and dialers) to a partial outsourcing (only the dialers, for instance).

10.3.2.1 Hosted predictive dialers

Hosted predictive dialing is a service provided by third parties that connects the calls made by the dialers of the service provider to the agents located within the company premises. The charging model of hosted predictive dialers uses the Software as a Service (SaaS) logic to provide organizations with a predictive-dialer capability.

Typically, the only requirement for an organization to use a hosted predictive-dialer system is a computer with a data connection and telephone lines for each agent or, when the host uses VoIP service for calls, a computer with a data connection and an IP phone for each agent.

10.3.3 Trends

Two popular trends in predictive dialers are call blending and multimedia queuing—we will discuss both of them.

10.3.3.1 Call blending

The term "call blending" also known as "double skill" refers to the ability to "blend" inbound and outbound calls among the same group of agents, depending on the occupation. This means that if agents who typically handle inbound calls are idle for x amount of time, the system automatically links the dialer with the ACD and starts feeding outbound calls to these agents. But as soon as incoming call volume increases to the point at which more inbound agents are required, it will put them back on inbound calls. This process works the same way for agents that typically handle outbound calls.

The statistical models necessary to combine inbound and outbound operations has its difficulties to implement, and this strategy should be considered only if you have a clear pattern of low versus high traffic (inbound or outbound) during known periods of the day.

10.3.3.2 Multimedia call centers

A type of call center whose numbers tend to increase is what we call a "multimedia call center." These call centers have the ability to incorporate predictive dialing in operations that can handle customer transactions regardless of contact medium—for example, traditional phone calls, e-mail, Web-chat requests, or faxes.

In addition, dialers are evolving to take on a bigger set of operational attributes, such as cost of the call versus benefit and "best time to call," based on demographic and behavioral characteristics.

10.4   Factors influencing the performance of predictive dialers

In this section, we will address the factors affecting the performance of a predictive-dialing solution, including the connect rate generated, the ability to distinguish between an answering machine and a live-person response, and the lead-base penetration rates accomplished. Other important aspects discussed are hardware and software related issues that could optimize a dialer's performance and subsequently impact a call center's productivity.

Predictive dialers are sophisticated automated systems that, just as the normal dialers, call a list of numbers and transfer the call over to an agent when a human responds. The difference between a normal dialer and a predictive one is the fact that the predictive dialer can adjust the dial rate based on a set of parameters:

- Waiting time: time that the call can wait to be transferred to the agent
- Call duration: average duration of the call
- Agents available: number of agents available
- Loss: loss due to answering machines, bad numbers, and busy lines

The deployment of predictive dialers tends to increase productivity in call centers, since being able to adjust carefully the dial rate to the available agents increases the productivity of the agents and avoids unnecessary calls that would otherwise be charged (nuisance calls).

Therefore, a predictive dialer adds productivity to the benefits already generated by the traditional dialers (autodialers/power-dialers): agents spend more time talking to people, and the time waiting for calls to go through goes down (hang-up goes down too), saving money with calls.

Implemented correctly, predictive dialers are an important ingredient in call-center productivity. However, the degree of actual productivity experienced depends to a large degree on design decisions and capabilities of the product platform. To achieve optimal performance, your predictive-dialing solution should accomplish high connect rates and accurate answering and fax machine detection capabilities, which depend on the accuracy of your solution's tone detection and call-transfer capabilities. The objective is to maximize the lead-base penetration parameter—the relationship between the total numbers dialed and actual number of people contacted.

The objective is to adjust the dialing rate to the available agents in order to guarantee the maximum occupation of these agents and at the same time avoid the loss of the calls transferred for the attendance group (silence or nuisance calls).

In this context, we have to keep in mind that the process of call classification (identification of valid numbers) generates only a multiplier factor in the process of identifying the adequate dialing rate. The defining variables are, in fact, the number of available agents and the average duration of the calls. Very often, discussion about the process of identifying bad numbers and the accuracy of such process takes away the focus from the basics.

Ideally, the waiting time should be zero (tw=0) and the loss after transference (to the agents) must not exceed 5 percent, tending to zero. The closer to zero the waiting time, the bigger the contribution of the dialer for the productivity of the operation.

Here it is worth mentioning that three parameters are basic when evaluating a predictive-dialer operation. If you don't control anything else, you must control at least these three parameters:

1) Total number of calls (this can be gotten from the billing system or from the telephone bill)
2) Total number of calls transferred to the agents (numbers dialed minus the numbers which for any reason were not transferred)
3) Total number of calls actually attended (total number of calls transferred minus the calls dropped before actual attendance)

These three basic parameters may be identified for the whole operation or in a per-campaign basis. These parameters allow us to see:

1) The correlation between total calls dialed versus number of effective calls attended

2) Correlation between total numbers dialed versus valid numbers
3) Correlation between calls transferred to the group and calls affectively attended

Therefore, through these parameters, it becomes possible to verify if there is any problem in the dialing process or in the adjustments between dialing/transference/attendance.

Here it is important to keep in mind that the dynamical adjustment of the dialing rate is the defining characteristic of a predictive dialer, and fixing parameters used to the calculation of dialing rate (number of available agents, call duration, loss, or even dial rate) may help reduce the processing demand but ultimately reduces the ability of the device to "predict" the right dial rate and consequently approximates it to the conventional autodialers (power-dialers).

If you are a telecom manager evaluating a dialer solution, be aware to ask the right questions. A specific piece of equipment may seem less demanding in terms of processing power and consequently may cost less.

However, what is hidden is the fact that the device may not do all the calculations or does them fixing some variables or having lower refresh variables frequency (refresh the variables and re-do the calculations in wider time intervals). All these strategies tend to reduce the required processing power of the dialer device, and consequently its cost, but also reduce its effectiveness calculating the proper dialing rate, thereby causing loss of productivity.

Example: If you identify that there is a high percentage of calls dropped after being transferred to the attendance group, it is highly probable that you have a problem with the dialing rate (another sign is a high waiting time which also usually implies in a high dropping rate). This phenomenon is a clear indicator that there is an asynchrony between the dialing rate and the availability of agents.

If you are changing your infrastructure from pure autodialers (power) to predictive dialers, the control of these variables is crucial to allow you to extract the most from the new devices.

The verification of the suggested parameters becomes even more effective if combined with the comparison between the growth of the number of calls and the growth of the number of attendants. A predictive dialer tends to produce more calls then the autodialers and tends to guarantee more time effectively talking to the agents—in other words, more productivity. But all depends on good configuring. A badly configured predictive dialer may be worse than a good power-dialer.

A company looking to deploy predictive dialers can either select a readily available turnkey solution or create a custom application using a development toolkit. Regardless of the chosen strategy, optimum performance requires careful planning. The factors to be verified when evaluating/planning a predictive-dialing solution:

- Connect rate
- Agent transfer time
- Answering-machine detection
- Fax machine detection
- Lead-base penetration parameters

We are going to discuss each one of these factors.

10.4.1.1 Connect Rate

The *connect rate* is defined as the total number of calls connected to the agents divided by the total number of calls dialed (it can be general or on a campaign basis or even on an agent basis). The connect rate is affected by several factors, including call-progress analysis (detection of bad numbers, busy tones, fax machines, and answering machines), call-transfer time, number of agents available, and call duration. The ability of a predictive dialer to correctly identify the call result it receives affects the number of appropriate calls made (dial rate).

A high-quality predictive dialer will accurately detect whether a live person has answered the call or whether it encountered a busy signal, operator intercept, or a fax or answering machine. Usually when an answering machine or a busy signal is detected, the number called is kept in the database for later contact. However, when an operator intercepts the call or a Special Information Tone (SIT) is detected, the number is discarded from the database, not to be called again. SIT tones are three, precise, sequential tones returned to a dialer when a connection cannot be made with a telephone number. The most common SIT tone is the vacant number intercept SIT ("The number you have dialed is not in service").

This tone detection is performed through call-progress analysis at the hardware level. Voice cards recognize call-progress tones based on frequencies and cadence (pattern of alternating silence and speech), which are predetermined and stored in a defined-tone table. During call-progress analysis, the voice card detects the frequencies and cadence and compares them to the patterns stored in the table to identify their meaning.

Since tone detection is usually hardware based, the selection of a telephony card can affect the dialer's performance. You must be sure that the card you choose or the card that comes with the prepackaged solution gives you the possibility to customize the call-progress analysis parameters. For example, the tones and the frequencies to be expected with each call-progress condition, such as busy, fast busy, disconnect, and no answer.

Telephony cards also differ in the number of seconds or cycles of tone pattern they listen to, before reporting the presence of a specific tone. The longer the card listens to the tones, the more accurate the detection. However, you should make sure that your card reaches equilibrium between call-progress analysis time frame and accuracy.

The fine tuning of the signaling of the public network trunks is usually very demanding in terms of time and effort. Sometimes, the adjustments have to be made in a trunk-by-trunk basis. Very often, the cards themselves have limitations about the tones they are able to understand or treat. In such cases, you may have to define which ones occur more often and leave the less-frequent ones for the operator. A typical mistake here is to try to save money buying second-rate cards, which have too-small processing capacity or memory limitations, because you pay less one time but have operational problems forever.

Some of the functions provided by the telephony cards—in particular answering machine and fax detection—can also be performed by software, which allows you to disable the card's tone detection capability and performs the call-progress analysis directly. Although it varies depending from the quality of the mailing, typical values for overall *connect rate* usually spin around 20 percent to 35 percent.

10.4.1.2   Agent-transfer time

If the predictive dialer detects a human response, the transfer to an agent should be transparent to the receiver of the call—as explained in the Erlang C calculation section (see chapter 11). *Waiting time (tw)* must be zero or very close to zero. This means that there shouldn't be any delays or silence periods between the instant the call has been answered by a prospect and the time the agent has responded. This not only maintains a high quality of service and minimizes nuisance calls for the receiver (and saves money for the company by avoiding unnecessary calls), but in some countries, it is a matter of following what is defined by the law.

The predictive solution should thus minimize the transfer time as much as possible. That means adjust the dialing rate to the availability of the agents. The agent-transfer time has a direct impact on your connect rate since many individual recipients quickly realize that they are receiving a telemarketing call when there is a noticeable pause or silence period after they answer the call. This usually results in high hang-ups percentage (calls dropped waiting to be attended by agents), and consequently limited lead-base penetration.

10.4.1.3   Answering-machine detection

Although the *connect rate* has a significant impact on campaign results, it is the appropriateness of the connection that will ultimately determine whether the transferred call will enhance an agent's productivity. Calls answered by a live person but incorrectly classified as answering-machine responses—and thus not transferred—cause call abandonment and eliminate a potential sale from the calls transferred to an agent.

Here it is important to mention that the "call abandonment," "hang-ups," or "call drops" are terms that refer only to the calls in which a recipient terminates a call after the dialer has determined that a live contact answered the call. Therefore, a call erroneously classified as answered by an answering machine and not even transferred is not counted as "abandonment," "hang-up," or "dropped," thus distorting your statistics, and on the other hand, the incorrect transfer of an answering machine to an agent also reduces the productivity.

To decide whether they are connected to an answering machine or a live person and to potentially pass the call to an agent, predictive dialers usually need to hear a response from the call recipient. Usually, a short "Hello" means that a live person has answered the call, whereas a lengthy response signifies an answering machine. Your predictive-dialing application or voice card should enable you to measure the duration of the response. Your answering-machine detection strategy will be based on this duration. For instance, if the response received is long, the dialer should categorize it as an answering machine and assign the phone number to a call-back list.

Ideally, the duration of continuous speech for answering-machine detection purposes should be customizable by the user and measured in milliseconds. The longer the time frame set for continuous speech, the more accurate the detection process. In case silence is detected after a short utterance, the dialer should interpret it as a human response, and immediately transfer the call to an agent. This increases the accuracy of answering-machine detection, yet at the same time minimizes the transfer time to an agent when a person answers the call.

Software-based detection algorithms represent a faster alternative to hardware-based solutions. However, they sometimes require more configuration and testing efforts. Make sure that your system provides the flexibility to use both alternatives.

An effective dialer minimizes *call-abandonment* (also known as hang-ups or call drop) rates by not only ensuring agent availability before it places a call but also by classifying correctly the calls. The difference between a good dialer, which identifies correctly answering machines, and a bad one, can be huge.

10.4.1.4 Fax-machine detection

Similar to the answering-machine detection capabilities mentioned before, the dialer also has to be able to recognize fax machines. When fax-machine tones are detected, the dialer should be able to notify and register the result of the call, enabling the call-center planners to decide whether to delete the number from the dialing list or add it to a fax list. This enables the correct categorization of leads in the database.

The precision of the detection of fax machines is very important. Commonly, predictive-dialing applications discard numbers that have been recognized as fax machines. Valid telephone numbers may be deleted from your database because the predictive-dialing application has wrongly identified them as fax-machine responses, and vice versa—numbers that should have been detected as fax machines may be retained in the database because they have been classified as "no answer" or "answering machine."

10.4.1.5 Lead-base penetration parameters

Lead-base penetration rates accomplished are defined as the total number of completed calls, bad numbers, wrong numbers, sales, and refusals divided by the total amount of leads available for calling. Some examples of lead-base penetration parameters are:

- Total number of connected calls/total number of leads available = Parameter 1
- Bad numbers/total number of leads available = Parameter 2
- Answering machines/total number of leads available = Parameter 3
- Fax machines/total number of leads available = Parameter 4
- Hang-ups/total number of leads available = Parameter 5

Note that the sum of the five parameters has to be equal to 100 percent. We can also subdivide the number of connected calls (Parameter 1) into subgroups:

- Sales/total number of leads available = Parameter 6
- Refusals/total number of leads available = Parameter 7
- Wrong numbers/total number of leads available = Parameter 8

These parameters allow you to understand exactly what is going on with your dial list (also known as lead-base or mailing) and understand if anything unusual is happening.

Ideally, you should be able to see how these parameters behave along the day in real-time through graphics like the one shown below (as the mailing is being dialed):

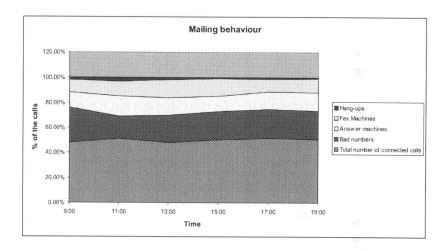

Although the values of each one of these variables are heavily dependent on the quality of the mailing deployed, things like the correlation between hang-ups and the total number of leads (Parameter 5) may indicate problems with the synchrony between the dialing rate and agents available, or an unusual high correlation between bad numbers and the total number of leads (Parameter 2) may indicate a problem when identifying SIT tones.

In the same token, an atypical correlation between fax machines/answering machines and the total number of leads (Parameter 3 and Parameter 4) may indicate a bad mailing but also can indicate problems with fax and answering-machine detection.

10.5   Preparing a dialer specification (RFP)

It is important to understand your specific needs in order to prepare an adequate RFP that guarantees you get what you need for the right price. Therefore, you need to define your dialing requirements, and based on them, identify the right technology and ideal contracting strategy. Only after that can you search for the adequate vendor. In other words, you can write the RFP only after you have defined your goals clearly. It is advisable to use a systems approach that encompasses:

Analysis→ Design→ Implementation→ Evaluation

When writing the RFP, you should be aware that vendors sometimes package their products and services in such a way as to make it difficult to make an easy comparison. For example, some hosted-dialer companies charge by the minute, others separate their charges by dialer port and local/long distance or VoIP minutes, and so on. In addition, vendors usually combine features and functions differently. Some charge more but include more functions, others break out and charge for each individual module or feature. The RFP is your instrument to define what you need and how you want to pay for it (trying to force some sort of the standardization in the proposals). It also helps to avoid paying for unnecessary features or functions.

Here we divided the process of preparing an RFP into four topics, as follows:

- Selecting the technical requirements
- Selecting the contracting strategy (premise versus hosted)
- Making sure that the specification exposes the capability, credibility, and longevity of potential vendors
- General recommendations

10.5.1 Defining the technical requirements

In order to adequately define the technical requirements you have to define the type of dialer most suitable for your specific operation, evaluate the need for integration with the CRM application to be deployed (the same already deployed by the organization or a new one), and identify the general features of the system.

10.5.1.1 Selecting the type of dialer

You have to decide if you need a conventional power-dialer or a predictive one, and which additional features your system has to have, would be good to have, and could have.

As mention before, power-dialers are best suitable for telesales, insides-sales or outside-sales organizations in which the sales process is more complex and requires multiple calls for closure. Predictive dialers, on the other hand, are the best alternative when volume of calls and time on the phone are the main requirements. Predictive dialers are more suitable for business-to-consumer applications with short and consistent lengths of calls and direct-dial phone numbers.

10.5.1.2 Conventional dialers (power-dialers)

Conventional dialers (also known as power-dialers or autodialers) execute the basic function of dialing at a constant rate and transferring the call to a human attendant. The dialer device can also perform several other tasks, such as announcing verbal messages, leaving messages on answering machines, or transmitting digital data (like SMS messages). There are typically three types of power-dialers:

- Conventional
- Voice-message dialer
- Click-to-call dialer

Conventional: This type of dialer executes the basic functions of dialing from a list and transferring the call to a human attendant. In addition, these devices usually can monitor the dialed numbers and change them to seamlessly provide services such as least-cost routing.

Voice-message dialer: This type of dialer is basically a conventional dialer that automatically dials a list of numbers and detects a live answer or fax/answering machine and plays a prerecorded voice message at the appropriate time. This is often called a voice-messaging dialer, or voice-message broadcasting. It completely automates the dialing process and is able to play a prerecorded message to hundreds or thousands of people in a short period of time.

Click-to-call dialer: The agent sets the pace, and the function is minimal. This dialer provides very little leverage in that all it does is save the time to dial the phone number from a list.

10.5.1.3 Predictive dialers

In addition to automatically dialing a list of phone numbers as conventional dialers do, the predictive dialers screen out no-answers, busy signals, special-information tones (SITs), and fax/answering machines, only sending calls that reach a live person to the agents and, most importantly, adjusting the dialing rate to match the number of available agents. The conventional dialer just dials at a constant rate.

10.5.1.4 CRM integration capability/API/Web services

Making an analogy, a dialer is a machine gun of calls, and the CRM database is the ammo depot. Therefore it is important to make sure that the correct agent is calling the list of names or leads that are most appropriate to his or her skills or location as fast as possible. This is the reason why one of the fundamental questions to ask when specifying a dialer solution is if the organization requires the integration of the dialer solution with an existing or future CRM.

It is important to know if the dialer system comes with a list or lead-management database (or integration with an existing one is required) and offers an API or Web services capability to easily move data in and out as required.

Therefore, when writing the RFP, it is crucial to make sure that all requirements linked with the integration with the support application are very well defined. In addition to clearly stating your needs, you must ask the potential vendors about their experience integrating their products with other clients and ask for references (and, of course, check those references).

10.5.1.5   General features

After deciding which type of dialer is most suitable for your specific operation and evaluating the requirements regarding the integration with the exiting (or future) CRM system, you still have to define the features that you consider necessary to your operation. We are going to list and comment on the main features to be evaluated:

- *Data import capabilities:* You have to define what is considered the minimal acceptable data importation. Keep in mind that the mailings will have to be imported to the dialers on a regular basis, therefore simplicity and a friendly process are very important. (Remember that the longer the time to load the data the longer the time out of action; depending on the type of operation, that may be significant.)
- *Real-time statistics and reports:* You should be able to get real-time and detailed statistics about your campaigns in progress and be able to get a variety of reports that help track specifics about particular accounts and campaign results. Therefore, your specification has to indicate which real-time statistics you have to have and which ones you would like to have. It is very advisable to include all eight lead-base penetration parameters described previously and trunks, cards, and agent status (and utilization) as minimal requirements in your monitoring tool and be able to see those parameters in a per-campaign basis.
- *Interface design and features:* You may need to define specific interfaces and may need specific information to be shown in the monitoring consoles. Every need has to be written, although most interfaces will cover most of the basics, and it is your job to make clear what you consider is minimally acceptable.
- *Fine tuning:* You should have some control over the precise timing issues surrounding predictive dialing, including how long agents have in order to wrap up after calls, maximum time between the time a human is found and an agent is switched through (tw = waiting time), how long a system should wait before redialing

numbers that didn't previously answer, and all timing parameters surrounding call-blending.

- *Flexible agent control*: You should be able to specify if you want some agents to be able to switch in between campaigns. For example, if a particular sales campaign has reached its overall sales target, you may wish to set rules that tell the dialer to switch agents.

- *Less-cost routing:* You should indicate your need for call routing, which can be a very important feature, although in large operations the ideal is to have a specific and dedicated device to do the call routing. Usually a tandem PBX which allows you to pool all the trunks of all providers.

- *Call-blending:* If you want to perform call-blending, the system will have to tightly integrate with your inbound switch. You may also check with your IVR manufacturer, since automated CT systems often front-end and work closely with a call center's switch.

- *No-call zones:* You should be able to feed your dialer "do not call" lists. These are lists of people who do not wish to be cold called. In some countries. legislation requires that you leave these people alone, as well as certain other rules (such as no phone calls after 9:00 p.m.).

- *Deployment of VoIP:* You have to decide how VoIP technologies will be deployed. You may decide to use only VoIP extension (including remote agent/telecommuting) or a combination of VoIP and conventional extensions (If your dialer is the device supporting the extensions). By the same token, you may have only VoIP trunks, only TDM trunks, or both.

- *Configurations features:* You have to state clearly how you expect to execute functions such as source tracking, real-time lead capture, skills definition, lead transfer capability, call-back event scheduling, data scrubbing, and data storage.

- *Structure features:* You may need to have your dialer also record the calls in addition to dialing (including the monitoring and recording storage), or you may need to have a client/server structure. These features must be clearly stated in your RFP requirements.

- *Call classifying:* You may define the techniques and strategies to be deployed to recognize and classify the several types of calls (answering machines, fax machines, PTT messages, etc.).

10.5.2   Defining the contract strategy

Once you have defined the technical requirements, it is time to evaluate the contracting modality most suitable for your specific context. Usually your decision will be between two possible alternatives (although a combination of both maybe possible): *premised-based solutions* **in which** you purchase the equipment, install the software, hook up phone and Internet lines, and maintain all of it on premise or *hosted solution* in which you "rent" access to a dialer installation through a data connection for a monthly or annual fee.

- *Premise dialers* typically require internal resources for maintenance and support. This strategy requires an up-front capital expenditure and the labor cost to maintain. Usually, this strategy is more appropriate for companies that already have a significant investment in IT and telecommunications support staff.
- *Hosted dialers,* on the other hand, are rapidly gaining popularity because of the payment model as well as the benefits of outsourcing to the provider all maintenance, upgrades, and professional services. This model requires very little if any staff to maintain.

The primary question to ask here concerns your business model and resources—if you are able to expend up-front capital and already have sunk costs in to IT resources. Additional questions may be worth asking: are your volumes seasonal, or do you want a variable-cost structure that grows incrementally? Both alternatives have pros and cons, and there is no predefined right strategy; the best alternative will depend on your specific context.

10.5.3   Evaluating potential vendors

When selecting the dialer provider, you should verify not only the vendor's technical capability but also its credibility within the market and its longevity (you are selecting a long-term partner). Therefore, be careful of talking only to the sales staff of a potential vendor. Be aware of vendors who are too slow to respond to your requests. Be especially careful of vendors who do not respond to each question or issue you bring up. This often indicates how they will treat you later. However, larger and more established vendors will typically have stronger processes but will be slower to respond yet surely will answer every question.

It is wise to keep two or three viable vendors in the process until the final selection. Sometimes this isn't possible because of technical capabilities, but it is still wise to do so for as long as possible. Never feel intimidated about asking the hard questions:

- How long have they been in business?
- Are they profitable?
- What background does their management team have?
- What are their service philosophies?
- How stable and redundant is their infrastructure if they are a hosted solution?
- How fast can they respond to troubles if they are a premise solution?
- Will they allow you to make a real-life test drive with full capabilities before you have to commit to a long-term relationship?

In addition to asking the questions, it is very advisable to adopt the following strategies during the selecting process:

- Run the dialer with your live data and lead sources with enough calls to get a good feel.
- Ask for on-site visits to a customer that is similar to your own proposed solution.
- Take the time to test different kinds of scenarios that exist within your own business.
- Get managers and users involved to make sure the experience from their point of view is satisfactory.
- Research the entire support package available from each vendor.
- Talk to people in their organization besides the sales teams.
- Talk to support and consulting personnel before you are committed.

- Look for companies that clearly define the boundaries of their capabilities—this is a much better approach than those who promise everything and can't deliver.
- Support personnel will typically give you more clear and direct responses than the sales staff. Ask how they will handle you after you have bought their serivces. Talk about upgrades, integrations, time frames, implementation, ongoing support, and escalation procedures. You should check references from current vendor customers in the areas of achieving deployment deadlines and responsiveness to on-site or remote support.
- Make sure you negotiate a trial period of thirty or sixty days in which you can opt out of an agreement if the vendor you have chosen has not been able to deliver up to your expectations. Keep in mind that you should execute the trials only with the providers short-listed; trials are a necessity but time consuming, both for the seller and for the buyer.

Be suspicious of a vendor who doesn't ask a lot of questions and who doesn't spend time with each area of your company that will be involved in the final solution.

Avoid asking for discounts in customization costs and consulting and implementation services—these "soft costs" often make a huge difference in the success of the project, and you may regret cutting the costs of this particular kind of support.

When researching the potential vendors, never neglect to research financials, press releases, and company trends. This kind of effort helps you stay away from vendors who are in financial trouble. Some of the best places to start checking out a potential vendor is by looking at their press releases and their financial performance.

Be ready to turn down a vendor if you uncover too many potential problems.

10.5.4 General recommendations

You should prepare the RPF carefully and be aware of the fact that even the best vendors will not be able to compensate a bad preparation on your part.

To be able to do the evaluation of the solution properly it is very important to define your baseline of performance for several months prior to installation of the solution. This allows you to compare your vendor performance and verify if the results are the ones desired. The ideal scenario is to measure a baseline of performance in terms of daily dials, call duration, hang-ups, agent availability/occupation, contacts, and qualifications for a sufficient time (three to six months) prior to the purchase. Only by doing that will you have a clear reference to check if the goals are being achieved.

Another important consideration is to be aware of the hidden costs. For instance, hosted solutions cost significantly less initially but may contain hidden costs in licensing agreements. Premise solutions may contain similar hidden costs and delays but in different places, such as the need to plan for your internal equipment hosting, power, backup, security, bandwidth, and fiber connections, which often cost a lot and take several months to plan and deliver.

Finally, you must be aware of the prices of the following items before you close the deal:

- Additional modules and functions
- Consulting and training
- Maintenance and upgrade policies, including services and support
- Custom reports
- Escalated service
- Emergency response
- Termination fees
- On-site visits and travel costs

Therefore, your RFP must specifically ask for quotations of such items.

Once again it is worth mentioning that the effort to buy a dialer solution is a once-in-a-while effort, but the potential benefits are reaped every day.

Chapter 11: Calculations

In this chapter we will address the calculations required to do a capacity planning and to understand the parameters of a predictive-dialing operation. The understanding of the theory involved in these calculations is important to allow you to calculate properly the necessary resources and to act wisely about the operational parameters when adjusting a predictive-dialing operation.

11.1   Calculations—Erlang calculators

Erlang is a nondimensional measurement unit used in traffic studies as a statistical measurement unit of traffic volume. The name "Erlang" comes from the Dane engineer A. K. Erlang, a pioneer in the study of telephone traffic and in the study of what is known today as the "queue theory." The traffic corresponding to a one Erlang refers to a one-resource (trunk) in continuous use for one hour.

For instance, if a given site has two telephone lines and both are in continuous use, that means two Erlangs of traffic. Other example: a radio channel occupied for thirty minutes during a hour supports 0.5 Erlangs of traffic.

Therefore, one Erlang can be considered as a utilization factor per unit of time. In the same manner, 100 percent of use of one trunk means one Erlang and 100 percent of use of two trunks means two Erlangs. For example, if a total usage of mobile phones in a given area is 180 minutes, that represents 180/60 = 3 Erlangs. In general, if the average frequency of calls entering is λ per time unit and if the average retention of calls is h, the traffic in Erlangs (A) will be as follows:

$A = \lambda h$

This type of calculation is used to define if a system is over-or under-dimensioned. For instance, if we measure traffic for one hour (during peak time), we may use this measurement to define the number of necessary active trunks in an E1 trunk. Here it is important to understand that the disparity between the number of Erlangs and the number of necessary trunks to transport the traffic is because the calls have an uneven distribution along the time. (One call doesn't finish, and another one initiates immediately.)

The traffic measured in Erlangs allows us to calculate the quality of service (QoS). There is a range of different formulas to execute this calculation. We present, therefore, the algorithms to calculate the necessary number of Erlangs based on denial of traffic (if the volume of traffic exceeds a given value, the caller receives a busy tone) and waiting (if the volume of traffic exceeds a given value, the caller is kept waiting until an attendant gets free). Such strategies are known by the names Erlang B and C respectively.

The algorithms described below identify, based on a given traffic in Erlangs, the necessary number of trunks. This calculation has as its basic input the traffic expressed in Erlangs and the quality of service desired. As already mentioned, the number of trunks necessary to support a given traffic can be calculated based on two premises: 1) if the traffic exceeds the volume it will be denied (user receives busy tone) or 2) by retention, a situation in which the user stays in line waiting for the next available attendant.

Based on denial of services—Erlang B

Our objective with this algorithm is to calculate the number of trunks necessary to handle a given amount of traffic (expressed in Erlangs), considering the calls distributed along a one-hour period. We have to define also the percentage of the calls tried that will be lost (denied). To identify the number of Erlangs, we have to divide the number of minutes handled during the hour by sixty. Therefore, A is the number of Erlangs identified (total minutes during the peak hour divided by sixty), Quality B is the expected loss, and N is the number of trunks necessary to handle this volume.

```
* PROGRAMA : ER1.PRG
* AUTOR : Luiz Augusto de Carvalho
* DATA : 22/02/2002
* OBJETIVO : Erlang B calculation

* N = Number of trunks
* A = Traffic in Erlangs
* B = loss

ERRO = 0.0001
*@ 06,01 clear to 25,119

N= erlan1 b=1 do while b>quality
A = erlan1
B = 1
I = 0
DO WHILE (I<N) .AND. (B>=ERRO)
I = I+1
X = A*B
B = X/(I+X)
* ? l,n,x,b
* wait
* IF I = N—1
* @ 13, 10 SAY "B"
* @ 13, 15 SAY B PICT '999.999999'
* @ 13, 40 SAY A
* ENDIF
ENDDO
* @ 14, 10 SAY "I = "
* @ 14, 15 SAY I
* @ 15, 10 SAY "B = "
* @ 15, 15 SAY B
N=N+1 enddo

RETURN
```

Based on call retention—Erlang C

The objective of this algorithm is to calculate the number of trunks necessary to support a given amount of incoming traffic in such way as to guarantee that the traffic will be able to be accepted, even if not attended to immediately. The Erlang C Calculation deals with ten parameters.

The first parameter needed is the average call-arrival rate. It doesn't matter what time unit is used to specify the arrival rate, as long as the same time unit is used for the average call duration.

(1) λ=Average arrival rate

The second factor to be specified is the average call duration. This must be expressed in the same time unit used for the call-arrival rate.

(2) Ts= Call duration

The third factor is the number of agents available.

(3) M=Number of agents

The term "traffic intensity" comes from the original application of Erlang C, which was for telephone networks, and the volume of calls was described as the "traffic." We need to calculate the traffic intensity as a preliminary step:

(4) Traffic Intensity $U=\lambda*Ts$

The agent occupancy, or utilization, is now calculated by dividing the traffic intensity by the number of agents (M). The agent occupancy will be between 0 and 1. If it is not less than 1, then the agents are overloaded and the Erlang C calculation is:

(5) Agent Occupancy $P=U/M$

Now we can calculate the main Erlang C formula. This formula looks complicated but is straightforward to calculate with a few lines of programming. The value of Ec(m,u) is needed to calculate the answers we actually want.

(6) Calculate the Erlang C formula:

$$E_C(m, u) = \frac{\dfrac{u^m}{m!}}{\dfrac{u^m}{m!} + (1-\rho) \displaystyle\sum_{k=0}^{m-1} \dfrac{u^k}{k!}}$$

Ec(m,u) is the probability that a call is not answered immediately, and has to wait. This is a probability between 0 and 1, and to express it as a percentage of calls we multiply by 100 percent.

(7) Probability a Call Waits Ec * 100

Having calculated Ec(m,u) it is quite easy to calculate the average waiting time for a call, which is often referred to as the "Average Speed of Answer—ASA" or "Average Waiting Time—Tw." We have to remember the time units we used for arrival rate and call.

(8) Average Speed of Answer or Average Waiting Time

$$T_W = \text{average waiting time} = \text{ASA} = \frac{E_C(m, u) \cdot T_S}{m \cdot (1-\rho)}$$

Where T is the average duration of the call in seconds.

Frequently, we want to calculate the probability that a call will be answered in less than a target waiting time. The formula for this is given here. Remember that, again, the probability will be on the scale 0 to 1 and should be multiplied.

(9) Calculated Service Level T = Target answer time

$$W(t) = Prob(\text{waiting time} \leq t) \quad = 1 - E_c(m, u) \cdot e^{-(m-u)\frac{t}{T_s}}$$

If the service level is specified and you want to calculate the number of agents needed, then you must do a bit of intelligent trial and error. You have to find the number of agents that will just achieve the service level you want. That means you should do interpolations until you find the combination that suits you. The algorithm below does it for you.

(10) Calculate Agents Needed

```
* PROGRAM : ErlangC.prg
* WRITTEN BY: Luiz Augusto de Carvalho
* DATE : 03/07/2002
* OBJETIVE : Calculates number of agents for a given traffic
********************************************************************
&& Number of agents
m=15
target = 15
do while m<10000
&& Ts= call duration in seconds
ts=240
&& average arrival rate in seconds
lambda=.2
&& U = traffic intensity
u=ts*lambda
&& agent ocupancy = U/M
P=u/m

A0=(u**m)
fatorial=1
for x=1 to m
 fatorial=fatorial*x
endfor
a=a0/fatorial

b4=(1-p)
b3=0
for k=0 to (m-1)
 b1=(u**k)
```

```
fatorial1=1
for x1=1 to k
fatorial1=fatorial1*x1
endfor
b2=b1/fatorial1
b3=b3+b2
endfor
b=b4*b3

E=a/(a+b) && probability that the call waits

tw=(e*ts)/(m*(1-p))
if abs(tw)<target
? "e=",e, "M=",m,"Ts=",ts,"Lamb=",lambda,"p=",p,"TW=",tw
&&wait
 exit
endif
m=m+1
enddo
```

11.2   Calculations in predictive dialers

Here it is interesting to explain how the previously described formulas apply to a predictive-dialer operation. First, we have to understand that, from the perspective of the agent's group, a predictive dialer must be calculated as we calculate an inbound traffic, with the difference that we don't expect to have any *waiting time (Tw),* and we expect to have an *agent occupancy (P)* as high as possible. Of course we have to feed the calculation with the *number of agents (M)* and *call duration (Ts),* which changes dynamically during the operation, and based on these parameters, we calculate the *average arrival rate (λ),* which is in fact the dialing rate of the device. Therefore, if we could put this process in a flow:

Predictive Dialer – operation logic

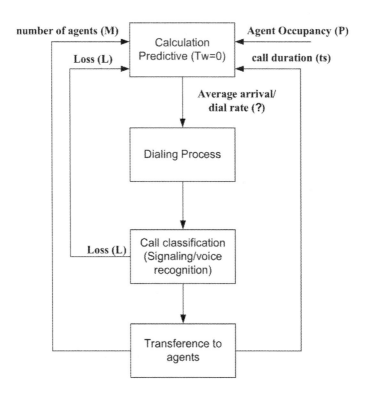

Note that the parameter Loss (L) doesn't appear in the formula. This parameter stands for the correlation between the numbers actually dialed and the numbers with success (bad numbers, answering machines, fax machines). This parameter works as adjuster in the *dialer rate* (λ).

The program below makes it possible to create a spreadsheet that correlates the parameters and shows how they vary.

```
* PROGRAM : ERLANG-C-DIALER.PRG
* AUTHOR : Luiz Augusto de Carvalho
* DATE : 22/02/2004
* OBJETIVE : Procedures with ERLANG C spreadsheets for dialers

&& Number of agents
m=15
target = 0
```

```
do while m<10000
&& Ts= call duration in seconds
ts=240
&& average arrival rate in seconds
lambda=0.01
do while lambda<5
&&lambda=.2
&& U = traffic intensity
u=ts*lambda
&& agent ocupancy = U/M
P=u/m

A0=(u**m)
fatorial=1
for x=1 to m
 fatorial=fatorial*x
endfor
a=a0/fatorial

b4=(1-p)
b3=0
for k=0 to (m-1)
 b1=(u**k)
 fatorial1=1
 for x1=1 to k
 fatorial1=fatorial1*x1
 endfor
 b2=b1/fatorial1
 b3=b3+b2
endfor
b=b4*b3

E=a/(a+b) && probability that the call waits

tw=(e*ts)/(m*(1-p))

erlangb=((lambda*60)*(ts/60)) &&minutes hour

* PROGRAMA : ER1.PRG
* AUTOR : Luiz Augusto de Carvalho
* DATA : 22/02/88
* OBJETIVE : Procedures with formula B de Erlang

dimension xc(1)

xc(1)=erlangb


&&? xc(1)
&&wait
```

Luiz Augusto de Carvalho and Olavo Alves Jr.

&&? "-----------------------------"


```
*PROCEDURE PERDAERL

* SUBROTINA PARA CALCULO DA PERDA EM UMA ROTA
* COM ACESSIBILIDADE PLENA
* N = Numero de Circuitos
* A = Trafego oferecido
* B = Perda calculada
for zz=1 to 1


ERRO = 0.0001
store 0.01 to quality
*@ 06,01 clear to 25,119

N= xc(zz)
store 1 to Bw
&&? "-",B,"-",quality
do while Bw>quality
Aw = xc(zz)
Bw = 1
I = 0

DO WHILE (I<N) .AND. (Bw>=ERRO)
I = I+1
X = Aw*Bw
Bw = X/(I+X)
* ? I,n,x,b
* wait
* IF I = N - 1
* @ 13, 10 SAY "B"
* @ 13, 15 SAY B PICT '999.999999'
* @ 13, 40 SAY A
* ENDIF
ENDDO
* @ 14, 10 SAY "I = "
* @ 14, 15 SAY I
* @ 15, 10 SAY "B = "
* @ 15, 15 SAY B
N=N+1
enddo
&&? xc(zz),int(n)
endfor
&&? int(n)
&&wait
```

```
&&? tw,target,p,n
&&wait
&&&&&&&&&&&&&&&&&&&&&&&&&&&&&&&&&&&&&&

if abs(tw)<=target .and. p>.5
 &&wait
 &&exit
 ? "e=",e,"M=",m,"Ts=",ts,"Lamb=",lambda,"p=",p,"TW=",tw,"Erlang - b: ",
erlangb,"Troncos = ",int(n)
endif
 lambda=lambda+0.01
 enddo

m=m+1

enddo
? "e=",e,"M=",m,"Ts=",ts,"Lamb=",lambda,"p=",p,"TW=",tw
```

11.3   Traffic matrixes

Traffic matrixes are fundamental instruments in understanding the traffic flows. Usually, we define the lines and columns of a traffic matrix as the area codes within a given country (or region as NANP—North America Numeric Plan); however, we also can use prefixes, cities, states, countries, or when analyzing internal traffic within an organization, we may use branches. When analyzing call-center traffic, area codes are usually the best choice.

Traffic matrixes may indicate the number of calls, number of minutes, and costs between the area codes (A to B and B to A). Here we give an example of a program to generate a traffic matrix by area code based on a list of calls:

```
* PROGRAM : MATRIX2.prg
* WRITTEN BY: Luiz Augusto de Carvalho
* DATE : Versao 24/11/2003
* OBJETIVE : Traffic matrix generator
*******************************************************************

SET DEFA TO C:\Projetos\geral
set talk off
```

```
SELECT 1
USE c:\projetos\geral\contamaster.dbf
SELECT 2
USE pontos
*****************************************************************
dime numerofan(200)
for t=1 to 200
numerofan(t)=0
endfor
select 1
go top
do while .not. eof()
store area1 to tarea1
store area2 to tarea2
select 2
go top
locate for val(tarea1)=val(area)
if .not. eof()
store recn() to tnumarea1 && numerofan(N)
endif
za=0
zza=0
&&? tnumarea1,area,tarea1
for t=1 to 200
&&? numerofan(t),tnumarea1
if numerofan(t)=tnumarea1
za=1
zza=zza+1
&&wait
endif
endfor
&&? z
if za=0
tya=0
for t=1 to 200
if numerofan(t)=0 .and. tya=0
tya=1
tyya=t
```

```
endif
endfor
if tya=1
numerofan(tyya)=tnumarea1
&&? numerofan(tyya),tnumarea1,area
endif
endif

&&wait
select 2
go top
locate for val(tarea2)=val(area)
if .not. eof()
store recn() to tnumarea2 && numerofan(N)
endif

z=0
zz=0
&&? tnumarea1,area,tarea1
for t=1 to 200
&&? numerofan(t),tnumarea1
if numerofan(t)=tnumarea2
z=1
zz=zz+1
&&wait
endif
endfor
&&? z
if z=0
ty=0
for t=1 to 200
if numerofan(t)=0 .and. ty=0
ty=1
tyy=t
endif
endfor
if ty=1
numerofan(tyy)=tnumarea2
```

```
&&? numerofan(tyy),tnumarea2,area
&&wait
endif
endif
select 1
skip
enddo
********************************************************************
&& Totalizing by area code -
Dime centroareamin(200,200)
dime centroareaval(200,200)
dime centroareaquant(200,200)
Dime centroareaminX(200,200)
dime centroareavalX(200,200)
dime centroareanome(503)
dime centroareacod(503)


for t=1 to 503
centroareanome(t)=0
centroareacod(t)=0
endfor

for t=1 to 200

for t1=1 to 200
centroareamin(t,t1)=0
centroareaval(t,t1)=0
centroareaquant(t,t1)=0
centroareaminX(t,t1)=0
centroareavalX(t,t1)=0
endfor
endfor
********************************************************************
select 2
go top

y=1
```

```
do while .not. eof()
store municipio to centroareanome(y)
store area to centroareacod(y)
skip
y=y+1
enddo
&&wait
select 1
GO TOP
DO WHILE .NOT. EOF()
STORE AREA1 TO TCENTRO1
STORE AREA2 TO TCENTRO2
STORE val(DURACAO) TO TDURACAO
STORE val(VALORservi) TO TVALOR
SELECT 2

GO TOP
LOCATE FOR val(AREA)=val(TCENTRO1)
if .not. eof()
store recno() to torigem
endif

GO TOP
LOCATE FOR val(AREA)=val(TCENTRO2)
if .not. eof()
store recno() to tdestino
endif
x1=0
x2=0
for t=1 to 200
if numerofan(t)=torigem
x1=t
endif
if numerofan(t)=tdestino
x2=t
endif
endfor
if x1<>0 .and. x2<>0
```

```
&&? torigem,tdestino,x1,x2,numerofan(x1),numerofan(x2)
store centroareamin(x1,x2)+tduracao to centroareamin(x1,x2)
store centroareaval(x1,x2)+tvalor to centroareaval(x1,x2)
store centroareaquant(x1,x2)+1 to centroareaquant(x1,x2)
endif
SELECT 1
&& ? RECNO()
SKIP
ENDDO




dimension h(503)
dimension g(503)
dimension i(503)
gheader=" "
hheader=" "
iheader=" "

for t=1 to 503
h(t)=""
g(t)=""
i(t)=""
endfor
*******************************************************************
* Mont the matrix1 - h(duration), matrix2 - g(value), Matrix3 - i(quantity)
*******************************************************************
for t = 1 to 200
select 2
if numerofan(t)<>0
go numerofan(t)
h(t)=""+area+" "+municipio+" "
g(t)=""+area+" "+municipio+" "
i(t)=""+area+" "+municipio+" "
hheader=hheader+area+" "
gheader=gheader+area+" "
```

```
iheader=iheader+area+" "
endif
endfor


*****************************************************************
* the lines
*****************************************************************
for t=1 to 200
for t1=1 to 200
store h(t)+str(centroareamin(t,t1),10,0)+" "to h(t)
store g(t)+str(centroareaval(t,t1),10,0)+" "to g(t)
store i(t)+str(centroareaquant(t,t1),10,0)+" "to i(t)
endfor
endfor
*****************************************************************
* matriz 1 - volume
*****************************************************************
use matrix1
dele all
pack
append blank
if len(hheader)<255
replace campo1 with substr(hheader,1,254)
else
if len(hheader)<509
replace campo1 with substr(hheader,1,254)
replace campo2 with substr(hheader,255,254)
else
if len(hheader)<763
replace campo1 with substr(hheader,1,254)
replace campo2 with substr(hheader,255,254)
replace campo3 with substr(hheader,509,254)
else
if len(hheader)<1017
replace campo1 with substr(hheader,1,254)
replace campo2 with substr(hheader,255,254)
replace campo3 with substr(hheader,509,254)
replace campo4 with substr(hheader,763,254)
```

```
else
if len(hheader)<1271
replace campo1 with substr(hheader,1,254)
replace campo2 with substr(hheader,255,254)
replace campo3 with substr(hheader,509,254)
replace campo4 with substr(hheader,763,254)
replace campo5 with substr(hheader,1017,254)
else
if len(hheader)<1525
replace campo1 with substr(hheader,1,254)
replace campo2 with substr(hheader,255,254)
replace campo3 with substr(hheader,509,254)
replace campo4 with substr(hheader,763,254)
replace campo5 with substr(hheader,1017,254)
replace campo6 with substr(hheader,1271,254)
else
if len(hheader)<1779
replace campo1 with substr(hheader,1,254)
replace campo2 with substr(hheader,255,254)
replace campo3 with substr(hheader,509,254)
replace campo4 with substr(hheader,763,254)
replace campo5 with substr(hheader,1017,254)
replace campo6 with substr(hheader,1271,254)
replace campo7 with substr(hheader,1525,254)
else
if len(hheader)<2033
replace campo1 with substr(hheader,1,254)
replace campo2 with substr(hheader,255,254)
replace campo3 with substr(hheader,509,254)
replace campo4 with substr(hheader,763,254)
replace campo5 with substr(hheader,1017,254)
replace campo6 with substr(hheader,1271,254)
replace campo7 with substr(hheader,1525,254)
replace campo8 with substr(hheader,1779,254)
else
replace campo1 with substr(hheader,1,254)
replace campo2 with substr(hheader,255,254)
replace campo3 with substr(hheader,509,254)
```

```
replace campo4 with substr(hheader,763,254)
replace campo5 with substr(hheader,1017,254)
replace campo6 with substr(hheader,1271,254)
replace campo7 with substr(hheader,1525,254)
replace campo8 with substr(hheader,1779,254)
replace campo9 with substr(hheader,2033,254)
endif
endif
endif
endif
endif
endif
endif
endif
********************************************************************
* matrix 1 - volume
********************************************************************
for t = 1 to 200

&& ? h(t)
append blank
if len(h(t))<255
replace campo1 with substr(h(t),1,254)
else
if len(h(t))<509
replace campo1 with substr(h(t),1,254)
replace campo2 with substr(h(t),255,254)
else
if len(h(t))<763
replace campo1 with substr(h(t),1,254)
replace campo2 with substr(h(t),255,254)
replace campo3 with substr(h(t),509,254)
else
if len(h(t))<1017
replace campo1 with substr(h(t),1,254)
replace campo2 with substr(h(t),255,254)
replace campo3 with substr(h(t),509,254)
replace campo4 with substr(h(t),763,254)
```

```
else
if len(h(t))<1271
replace campo1 with substr(h(t),1,254)
replace campo2 with substr(h(t),255,254)
replace campo3 with substr(h(t),509,254)
replace campo4 with substr(h(t),763,254)
replace campo5 with substr(h(t),1017,254)
else
if len(h(t))<1525
replace campo1 with substr(h(t),1,254)
replace campo2 with substr(h(t),255,254)
replace campo3 with substr(h(t),509,254)
replace campo4 with substr(h(t),763,254)
replace campo5 with substr(h(t),1017,254)
replace campo6 with substr(h(t),1271,254)
else
if len(h(t))<1779
replace campo1 with substr(h(t),1,254)
replace campo2 with substr(h(t),255,254)
replace campo3 with substr(h(t),509,254)
replace campo4 with substr(h(t),763,254)
replace campo5 with substr(h(t),1017,254)
replace campo6 with substr(h(t),1271,254)
replace campo7 with substr(h(t),1525,254)
else
if len(h(t))<2033
replace campo1 with substr(h(t),1,254)
replace campo2 with substr(h(t),255,254)
replace campo3 with substr(h(t),509,254)
replace campo4 with substr(h(t),763,254)
replace campo5 with substr(h(t),1017,254)
replace campo6 with substr(h(t),1271,254)
replace campo7 with substr(h(t),1525,254)
replace campo8 with substr(h(t),1779,254)
else
replace campo1 with substr(h(t),1,254)
replace campo2 with substr(h(t),255,254)
replace campo3 with substr(h(t),509,254)
```

```
replace campo4 with substr(h(t),763,254)
replace campo5 with substr(h(t),1017,254)
replace campo6 with substr(h(t),1271,254)
replace campo7 with substr(h(t),1525,254)
replace campo8 with substr(h(t),1779,254)
replace campo9 with substr(h(t),2033,254)
endif
endif
endif
endif
endif
endif
endif
endif
&&wait
endfor


copy to matrix1.txt sdf
********************************************************************
* matrix 2 - value
********************************************************************
*
&& wait
use matrix2
dele all
pack
append blank
if len(gheader)<255
replace campo1 with substr(gheader,1,254)
else
if len(gheader)<509
replace campo1 with substr(gheader,1,254)
replace campo2 with substr(gheader,255,254)
else
if len(gheader)<763
replace campo1 with substr(gheader,1,254)
replace campo2 with substr(gheader,255,254)
replace campo3 with substr(gheader,509,254)
```

```
else
if len(gheader)<1017
replace campo1 with substr(gheader,1,254)
replace campo2 with substr(gheader,255,254)
replace campo3 with substr(gheader,509,254)
replace campo4 with substr(gheader,763,254)
else
if len(gheader)<1271
replace campo1 with substr(gheader,1,254)
replace campo2 with substr(gheader,255,254)
replace campo3 with substr(gheader,509,254)
replace campo4 with substr(gheader,763,254)
replace campo5 with substr(gheader,1017,254)
else
if len(gheader)<1525
replace campo1 with substr(gheader,1,254)
replace campo2 with substr(gheader,255,254)
replace campo3 with substr(gheader,509,254)
replace campo4 with substr(gheader,763,254)
replace campo5 with substr(gheader,1017,254)
replace campo6 with substr(gheader,1271,254)
else
if len(gheader)<1779
replace campo1 with substr(gheader,1,254)
replace campo2 with substr(gheader,255,254)
replace campo3 with substr(gheader,509,254)
replace campo4 with substr(gheader,763,254)
replace campo5 with substr(gheader,1017,254)
replace campo6 with substr(gheader,1271,254)
replace campo7 with substr(gheader,1525,254)
else
if len(gheader)<2033
replace campo1 with substr(gheader,1,254)
replace campo2 with substr(gheader,255,254)
replace campo3 with substr(gheader,509,254)
replace campo4 with substr(gheader,763,254)
replace campo5 with substr(gheader,1017,254)
replace campo6 with substr(gheader,1271,254)
```

```
replace campo7 with substr(gheader,1525,254)
replace campo8 with substr(gheader,1779,254)
else
replace campo1 with substr(gheader,1,254)
replace campo2 with substr(gheader,255,254)
replace campo3 with substr(gheader,509,254)
replace campo4 with substr(gheader,763,254)
replace campo5 with substr(gheader,1017,254)
replace campo6 with substr(gheader,1271,254)
replace campo7 with substr(gheader,1525,254)
replace campo8 with substr(gheader,1779,254)
replace campo9 with substr(gheader,2033,254)
endif
endif
endif
endif
endif
endif
endif
endif
*********************************************************************
* matrix 2 - Cost
*********************************************************************
for t = 1 to 200

&& ? g(t)
append blank
if len(g(t))<255
replace campo1 with substr(g(t),1,254)
else
if len(g(t))<509
replace campo1 with substr(g(t),1,254)
replace campo2 with substr(g(t),255,254)
else
if len(g(t))<763
replace campo1 with substr(g(t),1,254)
replace campo2 with substr(g(t),255,254)
replace campo3 with substr(g(t),509,254)
```

```
else
if len(g(t))<1017
replace campo1 with substr(g(t),1,254)
replace campo2 with substr(g(t),255,254)
replace campo3 with substr(g(t),509,254)
replace campo4 with substr(g(t),763,254)
else
if len(g(t))<1271
replace campo1 with substr(g(t),1,254)
replace campo2 with substr(g(t),255,254)
replace campo3 with substr(g(t),509,254)
replace campo4 with substr(g(t),763,254)
replace campo5 with substr(g(t),1017,254)
else
if len(g(t))<1525
replace campo1 with substr(g(t),1,254)
replace campo2 with substr(g(t),255,254)
replace campo3 with substr(g(t),509,254)
replace campo4 with substr(g(t),763,254)
replace campo5 with substr(g(t),1017,254)
replace campo6 with substr(g(t),1271,254)
else
if len(g(t))<1779
replace campo1 with substr(g(t),1,254)
replace campo2 with substr(g(t),255,254)
replace campo3 with substr(g(t),509,254)
replace campo4 with substr(g(t),763,254)
replace campo5 with substr(g(t),1017,254)
replace campo6 with substr(g(t),1271,254)
replace campo7 with substr(g(t),1525,254)
else
if len(g(t))<2033
replace campo1 with substr(g(t),1,254)
replace campo2 with substr(g(t),255,254)
replace campo3 with substr(g(t),509,254)
replace campo4 with substr(g(t),763,254)
replace campo5 with substr(g(t),1017,254)
replace campo6 with substr(g(t),1271,254)
```

```
replace campo7 with substr(g(t),1525,254)
replace campo8 with substr(g(t),1779,254)
else
replace campo1 with substr(g(t),1,254)
replace campo2 with substr(g(t),255,254)
replace campo3 with substr(g(t),509,254)
replace campo4 with substr(g(t),763,254)
replace campo5 with substr(g(t),1017,254)
replace campo6 with substr(g(t),1271,254)
replace campo7 with substr(g(t),1525,254)
replace campo8 with substr(g(t),1779,254)
replace campo9 with substr(g(t),2033,254)
endif
endif
endif
endif
endif
endif
endif
endif
&&wait
endfor
copy to matrix2.txt sdf
*******************************************************************
* matrix 3 - quant
*******************************************************************
*
&& wait
use matrix3
dele all
```

This program generates three sets of traffic matrixes: by quantity of calls between the area codes, by number of minutes, and by cost.

Here we can see an example of a traffic matrix in terms of number of calls per area code:

er Minutes

Area	City	11	9838	966	277A	346	138	412	443	182	192	147	124	371	312
11	SAO PAULO	222,625.00	6,992.00	770.00	1,249.00	4,980.00	2,082.00	12,963.00	574.00	3,716.00	18,417.00	2,608.00	648.00	242.00	25,636.00
9838	CANDIDO MENDES	20.00	2,187.00	90.00	0.00	0.00	0.00	0.00	0.00	0.00	0.00	0.00	0.00	0.00	0.00
966	BACABAL	43.00	147.00	218.00	0.00	20.00	0.00	0.00	0.00	0.00	0.00	0.00	0.00	0.00	0.00
277A	AFONSO CLAUDIO	0.00	0.00	0.00	399.00	16.00	0.00	0.00	0.00	0.00	0.00	0.00	0.00	0.00	0.00
346	ARAXA	929.00	0.00	0.00	0.00	34,781.00	0.00	108.00	0.00	115.00	27.00	13.00	5.00	0.00	395.00
138	REGISTRO	2.00	0.00	0.00	0.00	0.00	0.00	0.00	0.00	0.00	0.00	0.00	0.00	0.00	0.00
412	CURITIBA	81.00	0.00	0.00	0.00	508.00	0.00	6,388.00	19.00	0.00	11.00	0.00	0.00	0.00	11.00
443	COLORADO	7.00	0.00	0.00	0.00	0.00	0.00	688.00	159.00	0.00	0.00	0.00	0.00	0.00	0.00
182	ARARAQUARA	96.00	0.00	0.00	0.00	7.00	0.00	0.00	0.00	367.00	34.00	0.00	1.00	0.00	0.00
192	CAMPINAS	39.00	0.00	0.00	0.00	0.00	1.00	7.00	0.00	1.00	6,250.00	39.00	13.00	0.00	1.00
147	AVARE	47.00	0.00	0.00	0.00	0.00	0.00	0.00	0.00	20.00	54.00	762.00	3.00	0.00	8.00
124	CARAGUATATUBA	1.00	0.00	0.00	0.00	0.00	0.00	0.00	0.00	0.00	0.00	0.00	0.00	0.00	0.00
371	ABAETE	0.00	0.00	0.00	0.00	0.00	0.00	0.00	0.00	0.00	0.00	0.00	0.00	0.00	0.00
312	BELO HORIZONTE	1,811.00	272.00	30.00	3.00	71.00	61.00	640.00	29.00	154.00	571.00	68.00	41.00	200.00	13,011.00
6636	ARENAPOLIS	12.00	0.00	0.00	0.00	0.00	0.00	0.00	0.00	0.00	0.00	0.00	0.00	0.00	0.00
6580	AGUA BOA	0.00	0.00	0.00	0.00	0.00	0.00	0.00	0.00	0.00	0.00	0.00	0.00	0.00	0.00
6737	AGUA CLARA	17.00	0.00	0.00	0.00	21.00	0.00	5.00	0.00	0.00	0.00	0.00	0.00	0.00	0.00

er value

Area	City	11	9838	966	277A	346	138	412	443	182	192	147	124	371	312
11	SAO PAULO	R$ 13,867	R$ 1,124	R$ 44	R$ 60	R$ 436	R$ 160	R$ 2,417	R$ 45	R$ 262	R$ 3,378	R$ 216	R$ 101	R$ 21	R$ 4,072
9838	CANDIDO MENDES	R$ 2	R$ 258	R$ 23	R$ 0	R$ 1	R$ 0	R$ 0	R$ 0	R$ 0	R$ 0	R$ 0	R$ 0	R$ 0	R$ 0
966	BACABAL	R$ 6	R$ 66	R$ 132	R$ 0	R$ 2	R$ 0	R$ 0	R$ 0	R$ 0	R$ 0	R$ 0	R$ 0	R$ 0	R$ 0
277A	AFONSO CLAUDIO	R$ 0	R$ 0	R$ 0	R$ 311	R$ 16	R$ 0	R$ 0	R$ 0	R$ 0	R$ 0	R$ 0	R$ 0	R$ 0	R$ 0
346	ARAXA	R$ 359	R$ 0	R$ 0	R$ 0	R$ 10,023	R$ 0	R$ 0	R$ 0	R$ 98	R$ 38	R$ 5	R$ 4	R$ 0	R$ 235
138	REGISTRO	R$ 0	R$ 0	R$ 0	R$ 0	R$ 0	R$ 0	R$ 0	R$ 0	R$ 0	R$ 0	R$ 0	R$ 0	R$ 0	R$ 0
412	CURITIBA	R$ 46	R$ 0	R$ 0	R$ 0	R$ 198	R$ 0	R$ 520	R$ 5	R$ 0	R$ 6	R$ 0	R$ 0	R$ 0	R$ 0
443	COLORADO	R$ 11	R$ 0	R$ 0	R$ 0	R$ 0	R$ 0	R$ 127	R$ 133	R$ 0	R$ 0	R$ 0	R$ 0	R$ 0	R$ 0
182	ARARAQUARA	R$ 33	R$ 0	R$ 0	R$ 0	R$ 8	R$ 0	R$ 0	R$ 0	R$ 231	R$ 90	R$ 0	R$ 0	R$ 0	R$ 0
192	CAMPINAS	R$ 33	R$ 0	R$ 0	R$ 0	R$ 0	R$ 1	R$ 5	R$ 0	R$ 3	R$ 766	R$ 75	R$ 16	R$ 0	R$ 1
147	AVARE	R$ 29	R$ 0	R$ 0	R$ 0	R$ 8	R$ 0	R$ 0	R$ 0	R$ 14	R$ 30	R$ 490	R$ 1	R$ 0	R$ 14
124	CARAGUATATUBA	R$ 0	R$ 0	R$ 0	R$ 0	R$ 0	R$ 0	R$ 0	R$ 0	R$ 0	R$ 0	R$ 0	R$ 0	R$ 0	R$ 0
371	ABAETE	R$ 0	R$ 0	R$ 0	R$ 0	R$ 0	R$ 0	R$ 0	R$ 0	R$ 0	R$ 0	R$ 0	R$ 0	R$ 0	R$ 0
312	BELO HORIZONTE	R$ 344	R$ 26	R$ 2	R$ 0	R$ 12	R$ 4	R$ 74	R$ 2	R$ 12	R$ 81	R$ 12	R$ 3	R$ 49	R$ 1,306
6636	ARENAPOLIS	R$ 7	R$ 0	R$ 0	R$ 0	R$ 0	R$ 0	R$ 0	R$ 0	R$ 0	R$ 0	R$ 0	R$ 0	R$ 0	R$ 0
6580	AGUA BOA	R$ 0	R$ 0	R$ 0	R$ 0	R$ 0	R$ 0	R$ 0	R$ 0	R$ 0	R$ 0	R$ 0	R$ 0	R$ 0	R$ 0
6737	AGUA CLARA	R$ 1	R$ 0	R$ 0	R$ 0	R$ 12	R$ 0	R$ 0	R$ 0	R$ 0	R$ 0	R$ 0	R$ 0	R$ 0	R$ 0

er number of calls

Area	City	11	9838	966	277A	346	138	412	443	182	192	147	124	371	312
11	SAO PAULO	121,702	4,167	738	699	3,191	976	5,884	393	1,804	9,861	1,104	491	78	18,625
9838	CANDIDO MENDES	8	791	43	0	2	0	0	0	0	0	0	0	0	0
966	BACABAL	13	56	118	0	9	0	0	0	0	0	0	0	0	0
277A	AFONSO CLAUDIO	0	0	0	134	1	0	0	0	0	0	0	0	0	0
346	ARAXA	486	0	0	0	17,305	0	19	0	88	5	7	3	0	94
138	REGISTRO	2	0	0	0	0	0	0	0	0	0	0	0	0	0
412	CURITIBA	31	0	0	0	160	0	2,199	4	0	4	0	0	0	2
443	COLORADO	2	0	0	0	0	0	363	63	0	0	0	0	0	0
182	ARARAQUARA	16	0	0	0	3	0	0	0	196	23	0	1	0	0
192	CAMPINAS	14	0	0	0	0	1	2	0	2	5,910	21	7	0	1
147	AVARE	22	0	0	0	2	0	0	0	11	14	279	3	0	1
124	CARAGUATATUBA	1	0	0	0	0	0	0	0	0	0	0	0	0	0
371	ABAETE	0	0	0	0	0	0	0	0	0	0	0	0	0	0
312	BELO HORIZONTE	693	72	12	3	22	27	209	12	46	169	22	8	72	4,524
6636	ARENAPOLIS	7	0	0	0	0	0	0	0	0	0	0	0	0	0

Chapter 12: Closing Words

We hope you have enjoyed reading this book and that it proves useful as a reference guide in the future. We tried to consolidate a large set of information in a coherent way, which we expect will help you when managing your telecom infrastructure. We hope we have achieved our goal of providing a useful tool through which the IT and telecom managers in large call centers can improve the efficiency of their infrastructure.

This book is an attempt to share our experiences, which encompassed long professional careers in roles as telecom managers, IT managers, hardware vendor representatives, telco representatives, and independent consultants. We know that many of the topics covered in the book may be viewed from a different perspective, depending on which hat you wear. We took the perspective of telecom managers, and this book was written mostly for them. However, hardware vendors and telco representatives may also benefit from this book by improving their understanding of the challenges faced by their clients.

Many of the topics discussed demand some previous context of understanding from the reader, and many of the opinions expressed may be somewhat controversial in their interpretation by the authors. We don't expect you to agree with all of our opinions. In addition, we are very aware that organizational realities may drive a politically based decision-making process, which takes away the Cartesian line of thought expressed in this book. However, we made an honest effort to put the important factors up front and the reasoning behind them.

No organization does everything right, and perfection is something only God is capable of; we understand that and recognize that most telecom managers are often too busy just keeping the lights on. Nevertheless, having a clear view of the ideal situation is always important and provides a framework to work within and to measure progress in the right direction.

Bibliography

Ahuja, Ravindra K., Thomas L. Magnanti, and James B. Orlin. *Network Flows Theory, Algorithms, and Applications*, Prentice Hall, 1993

Ahuja, Vijay. *Design and Analysis of Computer Communication Networks.* McGraw-Hill, 1982

Bayer, Michael. *CTI Solutions and Systems—How to Put Computer Telephony Integration to Work.* McGraw-Hill, 1997.

Black, Uyless D. *Data Communications and Distributed Networks.* Reston, 1993.

Brosnan, Michael, John Messina, and Ellen Block. *Telecommunications Expense Management*—Miller Freeman

Buckland, Lori and Dave Bengston. *Call Center Technology Demystified.* Call Center Press, ICMI Inc., 2004.

Cahn, Robert S. *Wide Area Network Design: Concepts and Tools for Optimization.* Morgan Kaufmann,1998.

Daskin, Mark S. *Network and Discrete Location Models, Algorithms and Applications.* Wiley Inter-Science, 1995.

Dawson, Keith. *Call Center Handbook: The Complete Guide to Starting, Running, and Improving Your Call Center.* CMP Books, 2004.

Dijkstra, E. W., "A Note on Two Problems in Connection with Graphs," *Numerische Mathematik,* I:269–271, 1959.

Frank, H. and W. Chou. *Topological Optimization of Computer Networks, Proceedings of the IEEE,* 60:1385–1397, 1972.

Harnett, Donald L. *Statistical Analysis for Business and Economics.* Addison Wesley, 1993.

Kershenbaum, Aaron. *Telecommunications Network Design Algorithms.* McGraw-Hill in Computer Science Series, 1993.

Leônidas Conceição Barroso. *Cálculo Numérico*—Harper & Row do Brasil

Monma, C. L., and D. L. Sheng, "Backbone Network Design and Performance Analysis: A Methodology for Packet Switching Networks," *IEEE J. Select Areas Communications*, 4:946–965, 1986.

Network Analysis Corporation, ARPANET, "Design, Operation, Management and Performance." New York, April 1973.

Sharma, Roshan L. *Network Topology Optimization—The Art and Science of Network Design.* VNR Computer Library, 1990.

"Software ARIETE." WANPOT. 2005.

"Software TRMS" Telecommunications Resources Management System. WANOPT. 2005

Strother, S. C., *Telecom Cost Management.* Arthech House, 2002.

Worbel, Leo A. *Disaster Recovery Planning For Telecommunications.* Artech House, 1990.

"Wide Area Network Methodology." WANPOT. 2006.